COMPUTER NETWORKS

NETWORK TOPOLOGIES

TYPES, PERFORMANCE IMPACT AND ADVANTAGES/DISADVANTAGES

COMPUTER NETWORKS

Additional books in this series can be found on Nova's website under the Series tab.

Additional e-books in this series can be found on Nova's website under the e-book tab.

COMPUTER NETWORKS

NETWORK TOPOLOGIES

TYPES, PERFORMANCE IMPACT AND ADVANTAGES/DISADVANTAGES

CASANDRA RENDELL
EDITOR

NOVINKA
New York

For permission to use material from this book please contact us:
Telephone 631-231-7269; Fax 631-231-8175
Web Site: http://www.novapublishers.com

Library of Congress Cataloging-in-Publication Data

ISBN: 978-1-62618-180-9

Published by Nova Science Publishers, Inc. † New York

CONTENTS

PREFACE

In this book, the authors discuss the types, performance impact and advantages/disadvantages of network topologies. Topics include the analysis of the topological structure of networks and related problems via simple rules; application of artificial neural networks in the diagnosis of coronary heart disease; network topology, signal function pathways and cell function; model-based risk analysis of complex networks; and network analysis of clinical, medical and molecular data in systems science.

Chapter 1 - It has been widely recognized that networks is a good tool for describing various artificial and natural systems. Thus, a natural problem arises: how to construct networks and further properly study its features. At present, it is still lack of basic laws for describing the quantitative relationships in biological networks, social networks, etc. Therefore, the construction method of networks should be different from that used to build up the mathematical models in the past based on the quantitative relationships in various scientific fields, such as physics and mechanics. In this chapter, the authors provide a new method to construct networks via simple rules and further study related problem. The main feature of the method is that simple basic rules for constructing networks are firstly proposed based on the current scientific understanding of the problems discussed, and then the networks is builded up through these rules, and further the rationality of the rules is evaluated by comparing the topological characteristics of the constructed networks with those measured from real networks, related problem is finally analyzed using the obtained reasonable network models. Two examples will be given to illustrate the research approach. One example is about the modeling of protein-protein interaction networks and the other example is about the positioning of wireless sensor networks.

Chapter 2 - The present research is aimed to develop an ANN diagnostic model for the coronary atherosclerosis and ischemia for patients after coronary angiography on the basis of genetic, clinical laboratory and instrumental examination data. The analysis of the correlation between the signs allowed us to choose the factors most closely connected with the diagnosis. Hierarchical clustering and correlation analysis were adapted to allocate typical fields of diagnostic factors. Various types of ANN topologies (MLP, SVM, PCA, and hybrid network) were analyzed; the authors have found that the models based on ANN with principal components analysis, and double-layer perceptron ANN optimized with genetic algorithms achieve the best diagnostic efficacy.

Chapter 3 - The living beings continuously receive, process and respond to a myriad of signals. To this aim several transduction pathways have evolved, giving rise an amazingly complex functional dialogue into and among cells. Modern analytical techniques (such as 2D electrophoresis, DNA microarray, protein chips, etc…) increased dramatically our knowledge about the molecules involved in these signaling events and allowed the discovery of several biochemical determinants of signal transduction. Despite scientific community expectations, the applicative benefit of this always increasing information to date are scarce. In the authors opinion, this is due to the intrinsic complexity of biological beings, from molecules to cells, from tissues to organs, that make necessary the adoption of the computational modeling strategies. In this context, the authors adopted a biological networks-based approach to study the topology of signal transmission pathways. In particular the authors have realized and statistically analyzed the main topological indexes of biological networks representing the molecular events involved in 11 pathways of relevant importance for human biology [smooth and striated muscle contraction, release cycle of six neurotransmitters, visual phototransduction (rods), sperm capacitation, insulin signaling pathway, p53 pathway, regulation of retinoblastoma protein (pRb), mitochondrial ATP metabolism, glucose metabolism, signaling events mediated by stem cell factor receptor c-Kit and the circadian clock]. As a result the authors have found that the number of molecules composing each network is likely due to the balancing of energetic cost of each molecule and the need of system stability. In addition, the authors demonstrated that all the examined networks showed the same scale free topology and small world behavior with a clearly non hierarchical structure, unlike other biological networks. In particular the clustering coefficient (i.e. the measure of how each network node tend to cluster to other ones) was virtually zero and it was unrelated to the number of links per node, the characteristic path length (the measure of how many links it is necessary to pass through to travel between two random selected nodes) was about 6 and the averaged number of link per node ranged between 2 and 2.8. From these findings, it is possible to conclude that all examined pathways share the same topology that confers them some important biological features such as robustness against random failure, controllability, and specificity and efficiency in signal transmission, which are the specific signature of signal transduction pathways.

Chapter 4 - *Model-based Risk Analysis* (MBRA) is a method and tool for analyzing complex system risk as defined by a network of components and their links to one another. Network risk is defined as the expected consequence across all nodes and links of the network. Network nodes and links are the fundamental assets represented in MBRA. Nodes can be anything – a building, tunnel, computer, water pump, reservoir, power substation, person, group, or telecommunication switching office. Links can be anything: a power line, familial relation, pipe, Internet fiber optic cable, or virtual dependency. Nodes and links have properties: values needed to compute risk as well as network properties such as number of connections, betweeness, and height.

MBRA employs fault trees to model multiple threats arrayed against single or multiple nodes; and expected utility theory (EUT) to compute overall network risk. In addition to calculating EUT risk, MBRA performs various modeling and simulation functions as follows:

1 Network properties: compute degree, betweeness, height, and contagiousness.
2 Objectives: optimize on risk, vulnerability, probability of failure, and consequence.
3 Resource allocation: distribute budgets across nodes and links such that one of the objectives listed in 2 is optimized.
4 Return on Investment Analysis: analyze changes in objectives versus investment in threat, prevention, and response capability.
5 Cascades: simulate cascade failures precipitated by a single failure.
6 Flow: simulate the loss of flows through the network due to a single failure.
7 Displays: calculate and display the exceedence probability generated by simulation of faults in the cascade and flow simulations.
8 Threat analysis: model threats versus nodes and links as a fault tree.
9 Game theory: simulate a 2-person game and find optimal allocations of threat, vulnerability, and response investments using Stackleberg optimization.

10 Exceedence probability analysis and probable maximum loss analysis (an approach favored by the insurance industry).

MBRA has been used since 2006 to analyze dozens of critical infrastructure systems of interest to homeland security. Applications have mainly come from students doing classroom projects. Experience has showed that MBRA can be extremely expressive and capable of analyzing complete systems of all kinds and sizes. Some of MBRA's limitations have become apparent: threat, vulnerability, and consequence decrease as an exponential function of investment – a model that may not represent reality. Additionally, cascade failure probability in some systems, such as a power grid collapse or flood, actually increases as a collapse is taking place. MBRA assumes likelihood of collapse is a constant. Finally, MBRA does not model dynamical systems with feedback loops. It assumes directional or bidirectional flows, but does not model dynamic feedback. MBRA is freely available at www.CHDS.us for Windows and Macintosh operating systems.

Chapter 5 - Cancer is a large class of diseases. Systems level analysis of complex diseases such as cancer requires the analysis of relationships between different types of clinical as well as molecular data. Chemotherapy, in addition to radiotherapy and surgery, is used to treat cancers. However, cancer drugs mostly target general processes, i.e. DNA synthesis, toinhibit cell division. A recent trend has been to use targeted chemotherapy, to more selectively treat specific types of cancer. Systems science is the analysis of the global organization of the relationships between groups, i.e. the different types of diseases and drugs, and is in contrast to traditional approaches of analyzing individual parts of a system. Such analyses have provided an understanding of the combined clinical and molecular aspects of diseases such as cancer. Networks of diseases, drugs, and genes showed unexpected connectivity for particular members, correlations between the connectivity with epidemiological features, etc. Network analysis of clinical, medical and molecular data is a promising branch of systems science.

In: Network Topologies
Editor: Casandra Rendell

ISBN: 978-1-62618-180-9

Chapter 1

ANALYZING THE TOPOLOGICAL STRUCTURE OF NETWORKS AND RELATED PROBLEMS VIA SIMPLE RULES

Shuiming Cai[1], Qinbin He[2] and Zengrong Liu[3,*]

[1]Faculty of Science, Jiangsu University, China

[2]Department of Mathematics, Taizhou University, China

[3]Institute of Systems Biology, Shanghai University, China

ABSTRACT

It has been widely recognized that networks is a good tool for describing various artificial and natural systems. Thus, a natural problem arises: how to construct networks and further properly study its features. At present, it is still lack of basic laws for describing the quantitative relationships in biological networks, social networks, etc. Therefore, the construction method of networks should be different from that used to build up the mathematical models in the past based on the quantitative relationships in various scientific fields, such as physics and mechanics. In this chapter, we provide a new method to construct networks via simple rules and further study related problem. The main feature of the method is that simple basic rules for constructing networks are firstly proposed based on the current scientific understanding of the problems discussed, and then the networks is builded up through these rules, and

* Email: zrongliu@126.com.

further the rationality of the rules is evaluated by comparing the topological characteristics of the constructed networks with those measured from real networks, related problem is finally analyzed using the obtained reasonable network models. Two examples will be given to illustrate the research approach. One example is about the modeling of protein-protein interaction networks and the other example is about the positioning of wireless sensor networks.

I. INTRODUCTION

In recent years, with the rapid development of information technology and science technology as well as the extensive cross-penetration among various disciplines, complex networks have become a hot research field and attracted much attention from the scientific communities. Actually, if using nodes or vertices to represent basic elements with certain dynamical characteristics and information systems, while edges or links to represent the relationship or connection of these basic elements, then a wide range of systems in nature and society can be described by models of complex networks consisting of nodes connected by edges. For example, the cell can be described as a complex network of chemicals connected by chemical reactions; the Internet is a complex network of routers and computers linked by various physical or wireless links; the World Wide Web (WWW) is an enormous virtual network of webpages connected by hyperlinks. Other examples include food webs, social networks, organizational networks, coauthorship and citation networks of scientists, neural networks, cellular and metabolic networks, protein-protein interaction networks, electronic power grids, etc., [1-3]. The ubiquity of complex networks in science and technology has naturally led to a set of common and important research problems. Among them, the most basic issues are how to characterize network anatomy since structure always affects function. For instance, the topology of social networks affects the spread of information and disease, and the topology of the power grid affects the robustness and stability of power transmission [1].

In the early days, the study of complex networks always adopted an implicit assumption that the interaction patterns among the individuals of the networks can be embedded onto a regular and perhaps universal structure such as a Euclidean lattice. However, a number of studies have suggested that most real-life networks have no apparent design principles, and so they cannot be described as regular graphs. In late 1950s, two Hungarian mathematicians Paul Erdos and Alfred Renyi proposed an Erdos-Renyi (ER) random graph model to describe a network with complex topology [4]. Their work had laid a foundation of the random network theory, followed by intensive studies in the next 40 years and even today. Random graphs have been proposed as the simplest and most straightforward realization of large-scale networks with no apparent design principles. But our intuition clearly indicates that many real-life complex networks are neither completely regular nor completely random. Diverse complex systems, such as the cell and the Internet, must display some organizing principles which should be at some level encoded in their topology as well [2]. Therefore, tools and measures to capture in quantitative terms the underlying organizing principles need to be developed.

In the past few years, the computerization of data acquisition in all fields and the availability of high computing power have led to the emergence of huge databases on the topology of various real networks. The public access to the huge amount of real data has in turn stimulated great interest in trying to uncover the generic properties of different kinds of complex networks [5]. As a result, many quantities and measures have been proposed to characterize the structural properties of networks in recent years [6, 7]. In this endeavor, two significant discoveries are the small-world effect and the scale-free feature of most real-life complex networks [8, 9]. In addition, many other interesting topological properties of complex networks have also been revealed, such as hierarchical modularity and disassortativity [10-12]. These empirical findings naturally lead to a problem: how to construct networks that can reproduce the known structural properties observed in real network, since the models proposed in mathematical graph theory turned out to be very far from the real needs. Moreover, modeling the structure of a complex network would also lead to a better knowledge of its evolutionary mechanisms and to a better cottoning on its dynamical and functional behavior [7].

Unfortunately, it is usually lack of clear quantitative relationships in the complex network modeling, but only some descriptive terms are given therein. This results in the construction method of networks should be different from that used to build up the mathematical models in the past based on the quantitative relationships in various scientific fields, such as physics and mechanics. Consequently, there exists a problem of how to give some simple rules to construct the network model based on these descriptive terms. The main purpose of this chapter is to provide some introduction and insights into this problem, with emphasis on how to extract some simple rules from these descriptive terms, and then to construct the network model via these simple rules for achieving the goal of modeling. Two examples will be given to illustrate the research approach. One example is about the modeling of protein interaction networks [13] and the other example is about the positioning of wireless sensor networks [14].

II. Some Basic Concepts

In this section, we shall first introduce some definitions and notations, and then provide a brief review of several important quantities used to describe the topology of a network.

Definitions and Notations

Graph theory is the natural framework for the exact mathematical treatment of complex networks and, formally, a complex network can be represented as a graph. An *undirected* graph G consists of two sets $\mathrm{N}(G)$ and $\mathrm{E}(G)$, such that $\mathrm{N}(G) \neq \varnothing$ and $\mathrm{E}(G)$ is a set of unordered pairs of elements of $\mathrm{N}(G)$. The elements of $\mathrm{N}(G)=\{n_1, n_2, \cdots, n_N\}$ are the *nodes (or vertices, or points)* of the graph G, while the elements of $\mathrm{E}(G)=\{e_1, e_2, \ldots, e_K\}$ are its *edges (or links, or lines).* The number of elements in $\mathrm{N}(G)$ and $\mathrm{E}(G)$ are denoted by N and K, respectively. A node is usually referred to by its order i in the set $\mathrm{N}(G)$. In a undirected graph, each of the edges is defined by a pair of nodes i and j, and is denoted as (i, j) or e_{ij}. The edge is said to be *incident* in nodes i and j, or to join the two nodes; the two nodes i and j are referred to as the *end-nodes* of edges (i, j). Two nodes joined by an edge are referred to as *adjacent* or *neighboring.* The *neighborhood* of a node i, henceforth represented as $O(i)$, corresponds to the set of nodes adjacent to node i. In the complex network literature, it is often assumed that no *loops*, i.e., edges from a node to itself, or *multiple edges*, i.e., couples of nodes connected by more than one edge, exsit. Graphs with either of these elements are called *multigraphs* [15, 16].

In this chapter, the emphasis will be on undirected graphs rather than multigraphs or directed graphs. For multigraphs and directed graphs, the reader is referred to the references [6, 7] and the books [15, 16].

Average Degree

The simplest and perhaps also the most important characteristic of a single node is its *degree*. The degree k_i of a node i is usually defined to be the total number of its connections. Thus, the larger the degree, the "more important" the node may be in a network [5]. The *average degree* $<k>$ of a network is the average of k_i for all vertices in the network, that is,

$$<k>=\frac{1}{N}\sum_i k_i.$$

In a sparse network, the average degree $<k>$ usually satisfies $<k>\propto \ln N$ [2].

Average Shortest Path Length

Shortest paths play an important role in the transport and communication within a network. Suppose one needs to send a data packet from one computer to another through the Internet: the geodesic provides an optimal path way, since one would achieve a fast transfer and save system resources [7]. Therefore, shortest paths have played an important role in the characterization of the internal structure of a network [7]. If one represents all the shortest path lengths of a network as a matrix D in which the entry d_{ij} is the length of the geodesic from node i to node j. The *average shortest path length* $<L>$ of the network, then, is defined as the mean of geodesic lengths over all pairs of nodes [8]:

$$<L>=\frac{1}{N(N-1)}\sum_{i\neq j} d_{ij},$$

where N is the total number of nodes in the network. Here, $<L>$ determines the effective "size" of a network, the most typical separation between two nodes therein. In a friendship network, for example, $<L>$ is the average number of friends existing in the shortest chain connecting two persons in the network [5]. It was an interesting discovery that the average path length of most real complex networks is relatively small, even in those cases where these kinds of networks have many fewer edges than a typical globally coupled network with an equal number of nodes. This smallness inferred the small-world effect, hence the name of small-world networks [5].

Average Clustering Coefficient

A common property of social networks is that cliques form, representing circles of friends or acquaintances in which every member knows every other member. This inherent tendency to clustering is quantified by the clustering coefficient, defined as the average fraction of pairs of neighbors of a node that are also neighbors of each other [5]. Suppose that a node i in the network has k_i edges, which connect this node to k_i other nodes. These nodes are all neighbors of node i. Clearly, at most $k_i(k_i-1)/2$ edges can exist among them, and this occurs when every neighbor of node i connected to every other neighbors of node i. The *clustering coefficient* C_i of node i is then defined as the ratio between the number E_i of edges that actually exist among these k_i nodes and the total possible number $k_i(k_i-1)/2$, namely,

$$C_i = \frac{2E_i}{k_i(k_i-1)}.$$

The *average clustering coefficient* $<C>$ of the network is then given by the average of C_i over all the nodes in the network:

$$<C> = \frac{1}{N}\sum_i C_i,$$

which characterizes the overall tendency of nodes to form clusters or groups. By definition, $0 \le C_i \le 1$, $0 \le <C> \le 1$; and $<C>=1$ if and only if the network is globally coupled, which means that every node in the network connects to every other node. In a completely random network consisting of N nodes, $<C> \propto 1/N$, which is very small as compared to most real networks. It has been found that most large-scale real networks have a tendency toward clustering, in the sense that their clustering coefficients are much greater than $O(1/N)$, although they are still significantly less than one (namely, far away from being globally connected) [5]. This, in turn, means that most real complex networks are not completely random. Therefore they should not be treated as completely random and fully coupled lattices alike.

Given the clustering coefficients of the nodes, the clustering coefficient can be expressed as a function of the degree of the nodes [6]:

$$C(k) = \frac{\sum_i C_i \, \delta(k_i - k)}{\sum_i \delta(k_i - k)},$$

where $\delta(\cdot)$ denotes Kronecker's delta function, that is, the average clustering coefficient of all nodes with k edges. For many real networks, this function has $C(k) \propto k^{-\upsilon}$. This behavior has been associated with a hierarchical structure of the network, with the exponent υ being called its hierarchical exponent [10, 11].

Degree Distribution

Not all nodes in a network have the same number of edges. The spread in the number of edges a node has, or node degree, over a network is characterized by a distribution function $P(k)$, which is the probability that a randomly selected node has exactly k edges. A regular lattice has a simple degree sequence because all the nodes have the same number of edges; and so a plot of the degree distribution contains a single sharp spike (delta distribution) [5]. Any randomness in the network will broaden the shape of this peak. In the limiting case of a completely random network, the degree sequence obeys the familiar Poisson distribution; and the shape of the Poisson distribution falls off exponentially, away from the peak value $<k>$. Because of this exponential decline, the probability of finding a node with k edges becomes negligibly small for $k >> <k>$. In the past few years, many empirical results showed that for most large-scale real networks the degree distribution deviates significantly from the Poisson distribution. In particular, for a number of networks, the degree distribution can be better described by a power law of the form

$$P(k) \propto k^{-\theta}.$$

This power-law distribution falls off more gradually than an exponential one, allowing for a few nodes of very large degree to exist. Because these power-laws are free of any characteristic scale, such a network with a power-law degree distribution is called a scale-free network [5]. A characteristic of this kind of network is the existence of hubs, i.e. nodes that are linked to a significant fraction of the total number of edges of the network.

Average Nearest Neighbors Degree

The degree distribution completely determines the statistical properties of uncorrelated networks. However, a large number of real networks are correlated in the sense that the probability that a node of degree k is connected to anther node of degree, say k', depends on k. In these cases, it is necessary to introduce the conditional probability $P(k'|k)$, being defined as the probability that an edge from a node of degree k [7]. $P(k'|k)$ characterizes formally the node degree correlations, but it is difficult to evaluate experimentally, especially for fat-tailed distributions, as a as a consequence of the finite network size and the resulting small sample of nodes with high degree [6]. This problem can be overcome by defining the average nearest neighbors degree of a node i as [17]

$$k_{nn,i} = \frac{1}{k_i} \sum_{j \in O(i)} k_j,$$

Then, one can calculate the average degree of the nearest neighbors of nodes with degree k, defined as $K_{nn}(k)$ [17],

$$K_{nn}(k) = \frac{\sum_i k_{nn,i}\, \delta(k_i - k)}{\sum_i \delta(k_i - k)},$$

to measure the degree correlations. If there are no correlations, $K_{nn}(k)$ is independent of k. When $K_{nn}(k)$ is an increasing function of k, nodes of high degree tend to connect with nodes of high degree, and the network is classified asassortative, whereas whenever $K_{nn}(k)$ is a decreasing function of k, nodes of high degree tend to connect with nodes of low degree, and the network is called disassortative [18]. Degree correlations are usually quantified by reporting the numerical value of the slope of $K_{nn}(k)$ as a function of k. Another way to determine the degree correlations is by considering the Pearson correlation coefficient of the degrees at both ends of the edges [18].

Many quantities and measures of complex networks have been proposed and investigated in depth in the last decades [6, 7]. The several concepts introduced above play a key role in the recent study and development of complex networks theory. Measuring these basic properties of a complex network is the first step toward understanding its structure. The next step, then, is to develop an evolution model with a topology of similar statistical properties, thereby obtaining a platform on which related problem can be studied. Our main goal here is to present two examples to illustrate how to construct network models via simple rules. To achieve this, we start with the first example: constructing protein-protein interaction networks that can reproduce the known topological properties of real protein-protein interaction networks at present based on duplication and divergence.

III. THE MODELING OF PROTEIN-PROTEIN INTERACTION NETWORKS

Biological processes, which contribute to the phenotypes of living cells, are wired by interaction networks of various cellular components such as proteins, DNA, RNA, metabolites, and small molecules [19]. The structures of a great deal of such networks, including protein–protein interaction, metabolic, signaling and transcription-regulatory networks, have been revealed due to the development of high-throughput data-collection technologies and new technology platforms [20]. Interestingly, none of these networks are independent; instead they form a ‘network of networks’ that is responsible for an organism’s form, rhythms, and functions. A major challenge of contemporary biology is to embark on an integrated theoretical and experimental programme to map out, understand and model in quantifiable terms the topological and dynamic properties of these complex biological networks [20].

As protein-protein interactions are central to most biological processes, the systematic identification of all protein interactions is considered a key strategy for uncovering the inner workings of a cell [21]. Consequently, a number of experimental and computational techniques have been developed to systematically determine both the potential and actual protein interactions in several model organisms. At present, the information of PINs at the whole-genome level is available from many organisms including *Saccharomyces cerevisiae* (bakers yeast) [22-24], *Caenorhabditis elegans* (worm) [25], *Drosophila melanogaster* (fruit fly) [26], and *Homo sapiens* (human) [27, 28]. The study of these PINs has revealed that they all have a similar, nontrivial architecture. First, PINs are sparse graphs, with a small average number of links [29]. Second, PINs are scale-free networks [20, 29, 30]. This means that there is no typical number of links per node; rather the distribution of the number of links (k) per node (P) decays as a power law, following $P(k) \propto k^{-\theta}$. In other words, there are many nodes with few links and a small but still significant number of nodes (hubs) with many links. Third, PINs have a small-world architecture [20,25,26,29]. This implies that, on the one hand, they are highly clustered, on the other hand, the average shortest path length in the network is almost as low as that for random networks [31]. Fourth, PINs show a hierarchical modularity structure [11, 21]. In a network showing a hierarchical structure, $C(k)$, the average cluster coefficient of k-degree nodes, decays as a power-law $C(k) \propto k^{-\upsilon}$ [10, 21]. This indicates that a node with a small number of links has a high average cluster coefficient and belongs to a small subnetwork in which all nodes are densely connected, while a hub has a low average cluster coefficient and links different subnetworks. Fifth, PINs show a disassortative structure, in which $K_{nn}(k)$ (" nn" represents "nearest neighbors"), the average degree among the neighbors of all k-degree nodes, follows $K_{nn}(k) \propto k^{-\mu}$ [12, 31]. Therefore, the connections between a hub and a low-degree node are favored, while those among hubs and low-degree nodes are suppressed [12, 32].

The power-law degree distributions have previously been observed in a number of naturally occurring graphs such as communication networks, web graphs, research citation networks, neural nets, etc [2, 3]. This property cannot be observed in the classical random graph models studied by Erdos and Renyi [4]. However, it is possible to generate graphs that satisfy this property by an iterative process that adds one new node to the graph at each step and meanwhile the new node is preferentially attached to some m of the existing nodes that already has many links [9]. Unfortunately, such a *preferential attachment* model does not capture the essence of the genome evolution and hence cannot be used to model PINs. From the point of view of natural selection and biological evolution, duplication and divergence are two dominant evolutionary forces in shaping biological networks such as gene regulatory networks and PINs [33, 34]. Duplication is the driving force for creating new genes in genomes: at least 50% of prokaryotic genes [35, 36] and over 90% of eukaryotic genes [37] are products of gene duplication, while divergence is the origin of generating function diversity. Recent work has shown that interaction networks constructed on the principle of duplication and divergence tend to exhibit such characteristics [31, 38-49]. However, to the best of our knowledge, most of previous theoretical models cannot illustrate the hierarchical modularity structure. In Ref. 46, the authors proposed a model for evolving networks by merging module, and showed that scale-free features and hierarchical modularity can be reproduced independent of network size. Furthermore, they modified the merging module model in Ref. 46 by introducing a fitness-driven preferential attachment mechanism, and showed that besides scale-free features and hierarchical modularity, disassortativity can also be exhibited by the model [47]. But the models presented in Refs. 46 and 47 took a module as a duplication unit and also assigned a variable of fitness to each node. It is difficult to correspond these mechanisms with the evolution processes in biology. Therefore, a more reasonable evolving model of PINs based on some plausible biological evolutionary mechanism needs to be developed.

Network Model and Result

Recently, we have proposed a simple evolution model that incorporates duplication, deletion, heterodimerization and addition, and shown that the model can display series of topological characteristics of real PINs, such as scale-free connectivity, small-world pattern, hierarchical modularity, and disassortativity [13]. Concretely, starting from a connected random network, by introducing a rule of small-preference duplication of a node, meaning that the probability of a node chosen to duplicate is inversely proportional to its degree, and subsequent symmetric divergence plus nonuniform heterodimerization based on some plausible mechanisms in biology, a simple network evolution model is constructed. In a word, the network model is made to evolve and grow according to the following rules. We start from a connected random network with N_0 nodes, and at every time step the following operations are preformed until the size of the network reaches N:

(i) *Small-preference duplication.* A node A with degree k_A of the existing network is chosen with the probability $P(A) = k_A^{-1} / \sum_j k_j^{-1}$, and is duplicated to generate a new node A', which is connected to all the neighbors of node A

(ii) *Deletion.* For each node j linked to node A and node A', we choose randomly one of the two links (A, j) or (A', j) and delete it with probability α.

(iii) *Nonuniform heterodimerization.* A new link between node A and node A' is established with probability β, which is proportional to the number (n_{cn}) of common neighbors shared by these nodes, i.e., $\beta = \beta_0 n_{cn}$ (β_0 is a constant). When $\beta > 1$, it is defined to be one.

(iv) *Addition.* New links (not previously present) between the newly generated node A' and all the rest of the nodes in the existing network are added with probability γ.

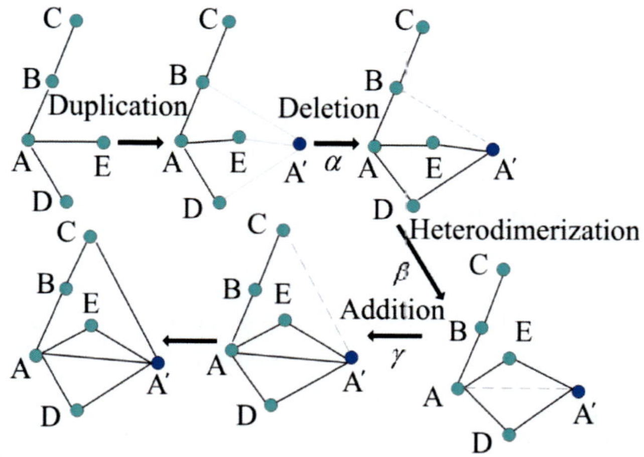

Figure 3.1. Schematic representation of the four operations in the proposed model: duplication, deletion, heterodimerization, and addition.

After this four-step procedure, if the new node A' is not connected to any other node, then it is removed. This guarantees the connectedness of the network [31, 44, 45, 48]. Different from random duplication of a node used in most of the exiting works, here it is assumed that the probability of duplicating a node in evolution is inversely proportional to its degree in our model based on the fact that core components of a PIN, such as the key nodes of high degree, tend to be conservative for their important functions in the network, but components at the periphery of the interactions are not [50]. Moreover, a negative correlation between the degrees of proteins and their duplicability was also reported in [19]. After the duplication, symmetric divergence, that is, links are lost from both of the duplicates with an equal probability is adopted here because it has been proposed that both duplication genes are subject to degenerative mutations and lose some functions, but jointly retain the full set of functions present in the ancestral gene [51]. In addition, in Ref. 31, the authors showed that a nonuniform heterodimerization model, in which a heterodimerization link is preferentially attached between duplicated nodes when they share many common neighbors, can successfully reproduce the high average cluster coefficient, the low number of heterodimerization link, and the skewed distribution of triangles demonstrated in the yeast PINs, which cannot be explained by a heterodimerization model where a new link is added between duplicated nodes with a uniform probability. Therefore, nonuniform heterodimerization is adopted in our model.

Without loss of generality, we take a single connected random network containing 100 nodes with the average degree $<k>=3.88$ as an initial network, and the final size of the network as 4000. The degree distributions of the network model under various parameter conditions are shown in Figure 3.2. The simulation results indicate that the degree distributions of the network model exhibits a power-law pattern in various parameter conditions, reflecting that the network has a scale-free feature.

Also, we can easily see that the power-law exponents (θ) of degree distribution of the different results that evolved are close to number 2.5, in accordance with the characteristics of real PINs [29, 30, 31]. In addition, we can observe that θ basically keep unchanging with increasing α in Figure 3.2(a), remain constant with increasing β_0 in Figure 3.2(b), and decay slightly with increasing γ in Figure 3.2(c). These results demonstrate that the scale-free feature in our model is robust to the model parameters as general biological networks do. This feature is determined mainly by the principle of small-preference duplication of a node.

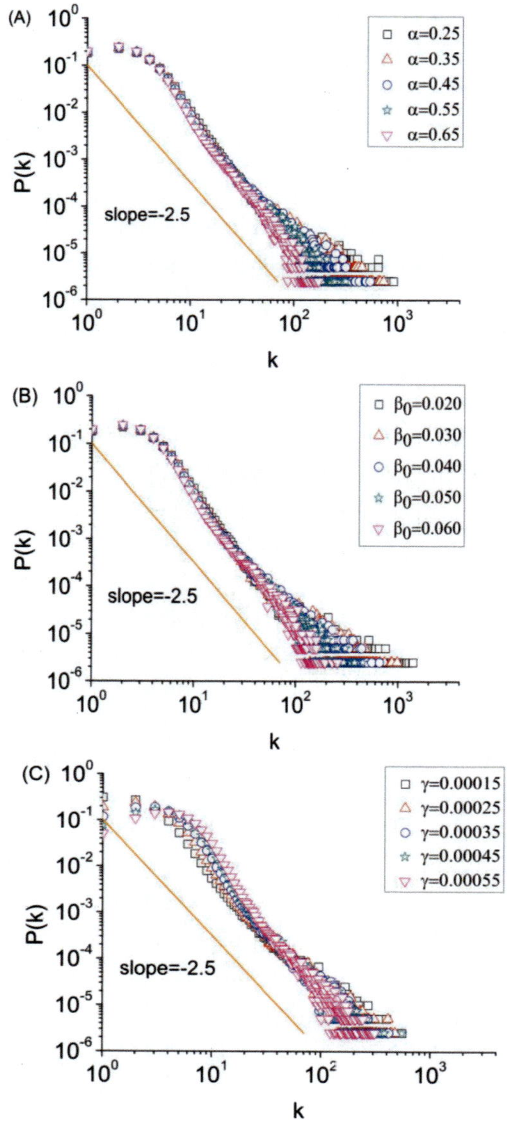

Figure 3.2. Degree distributions (log-log scale). (a) shows the degree distributions with $\beta_0 = 0.035$, $\gamma = 0.00025$, and α from 0.25 to 0.65; (b) shows the degree distributions with $\alpha = 0.45$, $\gamma = 0.00025$, and β_0 from 0.020 to 0.060; (c) shows the degree distributions with $\alpha = 0.45$, $\beta_0 = 0.040$, and γ from 0.00015 to 0.00055. The solid line (orange) of slope -2.5 is plotted for comparison. All the data are averaged over 100 independent runs.

Figure 3.3 shows the distributions of $C(k)$ of the network model under various parameter conditions. One can see that $C(k)$ versus k displays a flat head with a fat tail in the log-log plot, which is similar to the real observations and exhibits the hierarchical modularity [11]. It is well known that when the power-law exponent (υ) of the distribution of $C(k)$ is close to 1, the feature of hierarchical modularity is pronounced. Our simulation results indicate that an approximately inverse correlation between $C(k)$ and k is preserved at around $\alpha = 0.45$, $\beta_0 = 0.040$, and $\gamma = 0.00025$. In addition, we observe that power-law exponent υ mainly depends on the deletion probability α and the heterodimerization probability β_0. More concretely, υ grows with increasing α or β_0. The correlation of $K_{nn}(k)$ of the network model and k under various parameter conditions is shown in Figure 3.4. It can be seen that the decreasing trend of $K_{nn}(k)$ depending on k is displayed in various parameter conditions. This implies that the general patterns of disassortative mixing are generated by the network model regardless of the parameters. In addition, we can see that the power-law exponent (μ) of the distribution of $K_{nn}(k)$ mainly depends on the deletion probability α and the heterodimerization probability β_0 . More concretely, μ decays with increasing α or β_0 . Moreover, our simulation results indicate $K_{nn}(k) \propto k^{-0.5}$ for small k and $K_{nn}(k) \propto k^{-1}$ for large k at around α =0.45,
β_0=0.040, and γ=0.00025. The characteristic of $K_{nn}(k)$ has been verified by experiments, and the rational explanation has also been given [12]. Finally, the average degree $<k>$, the average cluster coefficient $<C>$, and the average shortest path length $<L>$ of the network model with various parameters are calculated, and the results are given in Table 1. We can see that these data agree well with the basic data of biological empirical networks [32]. Meanwhile, compared with the random networks with the same size, the average cluster coefficient $<C>$ of our network model is remarkably larger than that of the random networks. Therefore, the sparse and small-world properties are exhibited in this network model.

In summary, the above results show that the principle of small-preference duplication of a node may be the main driving force of the evolution of PINs.

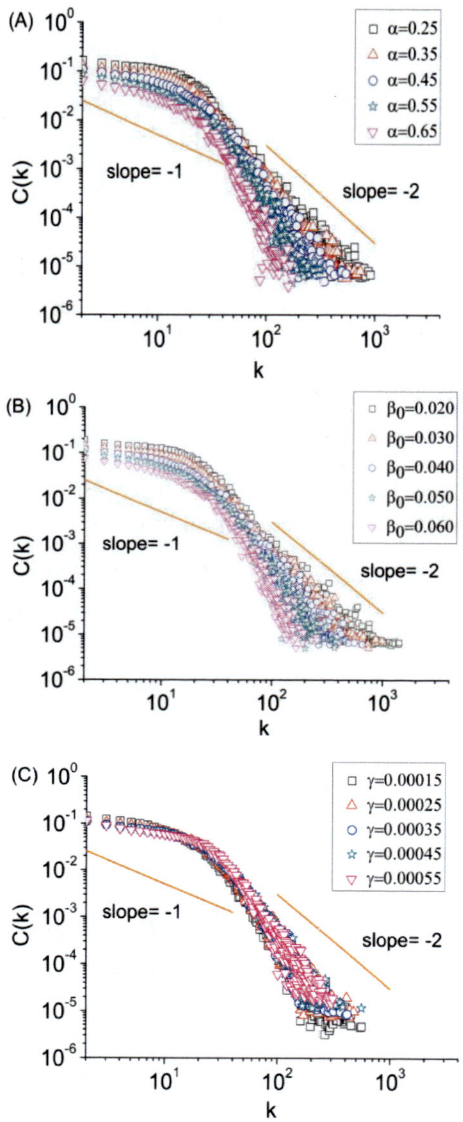

Figure 3.3. The distributions of $C(k)$ as a function k (log-log scale). (a) shows the distributions of $C(k)$ with $\beta_0 = 0.035$, $\gamma = 0.00025$, and α from 0.25 to 0.65; (b) shows the distributions of $C(k)$ with $\alpha = 0.45$, $\gamma = 0.00025$, and β_0 from 0.020 to 0.060; (c) shows the distributions of $C(k)$ with $\alpha = 0.45$, $\beta_0 = 0.040$, and γ from 0.00015 to 0.00055. The solid lines (orange) of slope -1 and slope -2 are plotted for comparison respectively. All the data are averaged over 100 independent runs.

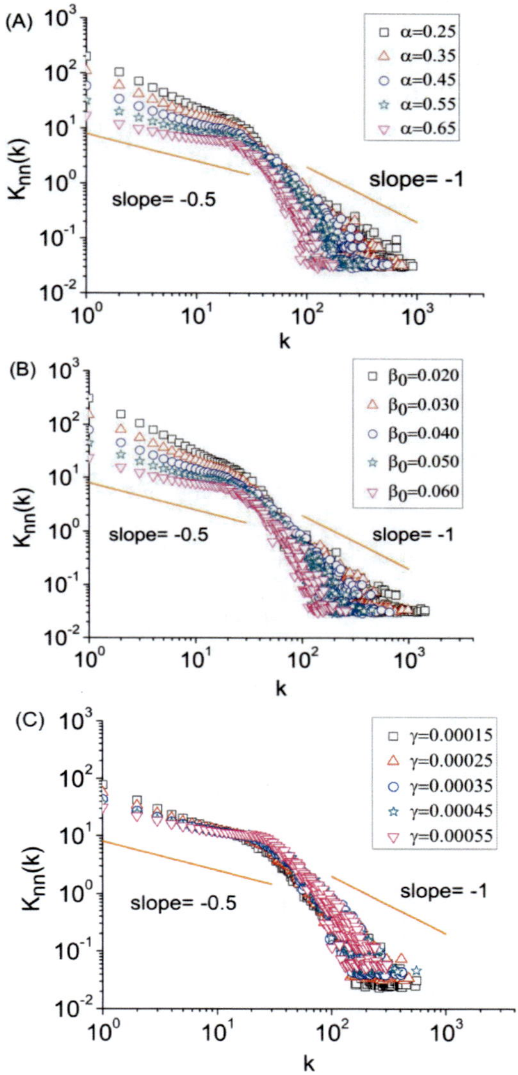

Figure 3.4. The distributions of $K_{nn}(k)$ as a function k (log-log scale). (a) shows the distributions of $K_{nn}(k)$ with $\beta_0 = 0.035$, $\gamma = 0.00025$, and α from 0.25 to 0.65; (b) shows the distributions of $K_{nn}(k)$ with $\alpha = 0.45$, $\gamma = 0.00025$, and β_0 from 0.020 to 0.060; (c) shows the distributions of $K_{nn}(k)$ with $\alpha = 0.45$, $\beta_0 = 0.040$, and γ from 0.00015 to 0.00055. The solid lines (orange) of slope -0.5 and slope -1 are plotted for comparison respectively. All the data are averaged over 100 independent runs.

Table 1. Values of the average degree $<k>$, the average cluster coefficient $<C>$, and the average shortest path length $<L>$ of the network model with different parameters α, β_0, and γ

Statistics	β_0=0.035 γ=0.00025					α=0.45 γ=0.00025				
	α=0.25	α=0.35	α=0.45	α=0.55	α=0.65	β_0 = 0.020	β_0 = 0.030	β_0=0.040	β_0=0.050	β_0=0.060
$<K>$	4.5842	4.3599	4.1294	3.8717	3.6238	4.7280	4.4904	4.2407	3.9996	3.7563
$<C>$	0.1111	0.0884	0.0717	0.0553	0.0410	0.1280	0.0998	0.0791	0.0643	0.0483
$<L>$	4.1740	4.5232	4.8766	5.2483	5.6781	4.8414	4.8385	4.8467	4.8631	4.9236

	α=0.45 β_0=0.040				
	γ=0.00015	γ=0.00025	γ=0.00035	γ=0.00045	γ=0.00055
$<K>$	3.5186	4.1598	4.7897	5.4339	6.0526
$<C>$	0.0828	0.0806	0.0758	0.0743	0.0712
$<L>$	4.8414	4.8504	4.8385	4.8766	4.8467

Three divergence mechanisms including deletion of links, addition of links, and heterodimerization have different effects on reproducing the statistical properties of real PINs. Especially, heterodimerization is of great importance in the emergence of hierarchical modularity. Based on these duplication and divergence mechanisms, our model can reproduce the basic statistical properties of known topological characteristics of real PINs. We believe that our results will enable us to deeply understand the evolutionary mechanism of PIN formation. Moreover, our results also show that the research approach proposed here, constructing network models via simple rules, is feasible.

Finally, it worth mention that although the duplication-divergence network model is generally thought to incorporate key ingredients underlying the growth and evolution of protein-protein interaction networks, properties of the model has always been elucidated through numerous simulation studies. At present, a comprehensive theoretical study of the model is still lacking. Recently, we have derived analytic expressions for quantities describing key characteristics of the network- the average degree, the degree distribution, the clustering coefficient, and the neighbor connectivity-in a mean field theory of the model in the large limit [52]. We found that all four quantities obeyed power-laws even at moderate network ($N \approx 10^4$), except the degree distribution, which had an additional exponential factor.

Table 2. Mean-filed estimates of power-law exponents of characteristic network quantities of the model

Power law	Exponent
Ave. deg. $K_N \sim N^{\beta_K}$,	$\beta_K = \begin{cases} 1-2p_d, & p_d < \frac{1}{2}; \\ 0 \text{ (logarithmic)}, & p_d = \frac{1}{2}; \\ 0, & p_d > \frac{1}{2}. \end{cases}$
Deg. dist.[a] $P(k) \sim k^{-\gamma_P}$,	$\gamma_P = 1, \quad p_d \le 0.4329;$
	$1 < \gamma_P \le 2,\ 0.4329 < p_d \le 0.5;$
	$2 < \gamma_P \le 3,\ 0.5 < p_d \le 0.5858.$
Clust. coef. $C(k) \sim k^{-\gamma_C}$,	$\gamma_C = \begin{cases} 1, & \frac{1}{2} \le p_d < 1; \\ 2p_d, & 0 < p_d \le \frac{1}{2}. \end{cases} \quad (p_h > 0)$
Deg. corr. $K_{nn}(k) \sim k^{-\gamma_{nn}}$,	$\gamma_{nn} = \begin{cases} 0, & p_d \ge \frac{1}{2}; \\ 2p_d - 1, & p_d < \frac{1}{2}. \end{cases}$

[a] Also has an exponential factor finite N.

We observed that the exponent of this factor itself obeyed a power-law [52]. The results of our derivation of the mean-field approximations of power laws of the quantities, the average degree, the distribution, clustering coefficient, and neighbor connectivity, are summarized in Table 2 [52].

In this section, we have discussed the basic simple rules for constructing protein interaction networks. In the following section, we will propose a positioning algorithm of wireless sensor networks by simple rules, and show that high performance localization can be obtained through this algorithm.

IV. The Positioning of the Wireless Sensor Network

During the past decades, wireless sensor networks (WSNs) have attracted considerable attention because of their potential applications in many engineering fields, such as habitat monitoring, forest-fire monitoring, battle-field surveillance, aerospace and space exploration. Sensors are generally equipped with data processing and communication capabilities. The sensing circuitry measures parameters from the environment surrounding the sensor and transforms them into electric signals. Recent technological advances have enabled the development of low-cost, low-power, and multifunctional sensor devices.

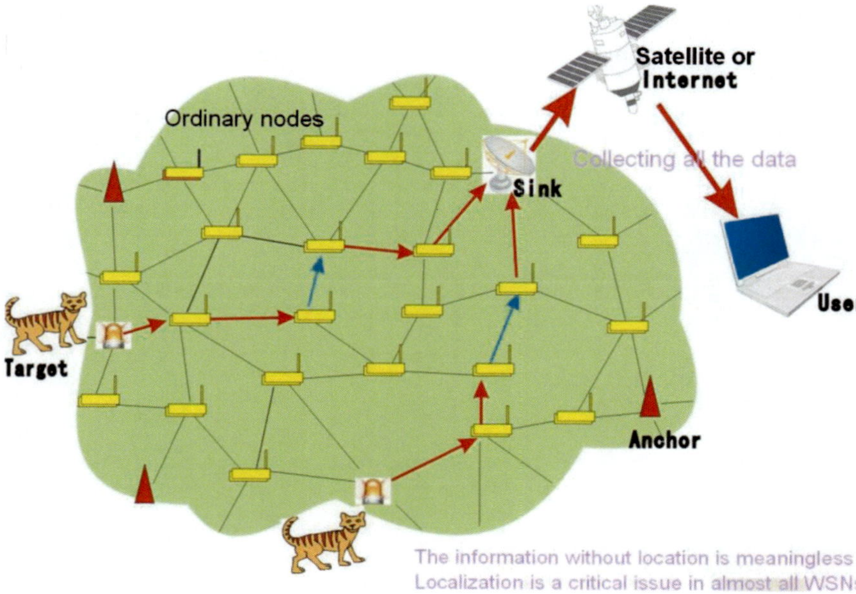

Figure 4.1. Structure and Function of WSNs.

Usually, the sensor nodes composed of WSNs can be divided into several types, such as ordinary nodes, anchor nodes (which are nodes with known their location information) and sink nodes (which are nodes collecting all the data). These nodes are autonomous devices with integrated sensing, processing, and communication capabilities.

To achieve information transmission, a wireless sensor network must a connectivity network. The information transmission of sensor network is via adjacent nodes communication and multi-hop transmission, as shown in Figure 4.1. These sensor nodes in a WSN are randomly placed and their exact locations may not be predetermined. It is obvious that the sensing data without knowing its location is meaningless. For rapid deployment in some complicated environments with a large number of low-cost sensor nodes, the use of GPS [53] or relatively expensive technologies is not practical [54–61]. Localization is a critical issue in almost all WSNs. The connectivity of a WSN is a necessary condition of positioning the network.

Usually, the localization algorithms can be divided into two classes: range-based and range-free. The range-based schemes need either node-to-node distances or angles for estimating locations. Range-based schemes use the time of arrival (TOA) [54], time difference of arrival (TDOA) [55–57], angle of arrival (AOA) [58, 59], and received signal strength indication (RSSI)

[60, 61]. The range-based schemes typically have higher positioning accuracy than the range-free schemes but require additional hardware to obtain the information of distances or angles.

The range-free schemes do not need the distance or angle information of sensor nodes for their localization [62–73]. The range-free schemes provide more economic and simpler estimates than the range-based ones. At present, the range-free schemes become more attractive than the range-based schemes. There are typical range-free schemes, such as DV-HOP [69,70], APIT [71], MDS-MAP [72], PDM [73]. The DV-HOP algorithm uses the hop-count from the corresponding anchors (nodes with known location information), the average-distance per hop and triangular positioning to obtain the location. The method requires that the node distribution is more dense and isotropic. To locate a WSN with average node degree 10, anchor ratio 10% and isotropic, the positioning accuracy of DV-Hop algorithm is 33% [69, 70]. APIT algorithm needs to determine if the unknown node is within a particular triangle formed by a set of anchors within radio range. Then, it calculates the centroid of this polygon area, and takes centroid as the location of unknown node. The algorithm requires that networks have strong connection strength. MDS-MAP algorithm can be applied to both range-free and range-based positioning. It requires considerable computation as a centralized algorithm and when the network node density decreases, the location error increases, and the number of unable to locate nodes increases. Experiment shows that as a range-free method to locate a WSN with average node degree 12.1, its positioning accuracy is about 30% [72]. PDM algorithm is SVD based method [73], which can be used in both isotropic and anisotropic WSNs. PDM algorithm achieves better positioning accuracy than MDS-MAP, especially in the case of anisotropic WSNs.

In addition, there are several optimization algorithms applied to localization WSNs, such as convex programming approach [68, 75], graph theoretic algorithm [76–78], genetic algorithm [79–81], and particle swarm optimization (PSO) [82].

However, almost all those existing methods have their various limitations, such as positioning accuracy, computational complexity, program convergence, hardware costs and error propagation issues.

Since the characteristics of information dissemination in a WSN is through local exchange, the positioning sensor network is also achieved by local communication. In order to achieve a certain function of a network, we often take the way of exchanging the information between the neighbor nodes.

This process is similar as Swarm and Flocking. Here, we introduce about Swarm and Flocking.

In a multi-agent system, we called Flocking that the agent be able to achieve equal speed and the stable distance between each other finally. The phenomenon of Flocking behavior is prevalent exist in the world, such as birds formation flight and Flocking phenomenon of fish [83, 84]. Flocking algorithm was first proposed by Reynolds in 1986. Three basic rules were proposed to describe Flocking: (1) separation; (2) cohesion; (3) alignment. Vicsek studied a simplified model of the Reynolds model in 1995. In the simplified mode, all of the agents maintain the same speed. In recent years, many control academics are research kinds of Flocking problems by constructing differential equations, and using artificial potential and kinetic energy method to achieve Flocking algorithm.

In addition, the characteristic of Swarm is composed of a collection of autonomous individual. Generally, the emergence of Swarm is perceived by individual local role and reaction behavior. The cluster phenomenon and cluster behavior is prevalent exist in the world in almost all scales, from the non-life world of molecules to galaxies, from the biosphere simple bacteria to higher animals. The research of Swarm has practical significance that is a way of understanding the biological complexity. On the one hand, you can learn from the wisdom of biological, and use the distributed strategy of control, coordination, and formation in autonomous multi-agent system (such as multi-agent robots or autonomous vehicle systems).

There are some general features of these systems, such as local communication and individual autonomy. Through the design of control algorithm, eventually, the system will show the emergence on the whole. It is shown the generation of swarm intelligence that each individual is the not very intelligent, but through collaboration between each individual and they can show a certain degree of intelligent behavior. A cluster is a distributed collaborative system, has a robust and self-organizing feature. Swarm intelligence based on local optimization system may be more advantageous than the traditional centralized control and more efficiency and robustness, and thus it has the engineering potential applications.

In Multi-Agent Systems, "consensus" means to reach an agreement regarding a certain quantity of interest that depends on the state of all agents. A "consensus algorithm" (or protocol) is an interaction rule that specifies the information exchange between an agent and all of its neighbors on the network [85]. The problem of consensus is mainly originated the issues of cooperative and control. The information sharing between agents is a prerequisite of the problem of cooperation and control for multi-agent systems. The sharing of information may appear in many forms, such as mutual control algorithm and mutual objective. When agents cooperate with each other to complete a task, effectiveness of cooperative control strategy make multi-agent deal with unpredictable situation and changes in the environment, which requires the agent adjust to the changes in the environment to achieve consistent. Multi-agent reach consensus is a precondition to achieve coordinated control. The consistency problem in computer science has a long history. In recent years, due to the extensive application of multi-agent systems, the problem of consensus is more and more under the attention of researchers.

As described above, we know that to establish some simple agreements between individual in network can make the whole network achieve a certain purpose. Thus inspired, we establish a simple agreement between the sensor nodes of wireless sensor networks to achieve network positioning. To positioning a WSN, some simple rules are proposed based on the idea of consistency problem (information dissemination by local communication). We propose a new range-free method for positioning WSNs, called spring swarm localization algorithm (SSLA).

SSLA and Examples

For a WSN, sensor nodes may be randomly distributed in a region. We consider a network in which a small fraction of sensors know their locations at all times. The sensor nodes known their locations are called anchors or beacons and their locations can be obtained by using a global positioning system (GPS), or by installing anchors at points with known coordinates. In addition, we assume senor nodes have the ability to know their own radio ranges.

Firstly, we present SSLA in detail as follows and we show that high performance localization can be obtained through this algorithm. The algorithm for localization is available for both isotropic and anisotropic WSNs. The positioning accuracy of the algorithm can reach 10% or even higher.

Here, consider a WSN with n nodes. Its position neighbouring graph is also denoted as $G=(\nu,\varepsilon)$, where $\nu=(node0, node1, \ldots, node(n-1))$ is a set of vertices and $\varepsilon=\{(nodei, nodej)\in \nu\times\nu : nodej \sim nodei\}$ represents a set of arcs, these arcs describe the position neighbouring relations of unordered pairs of vertices. And, denote $x_i = (x_{i1},..., x_{im})^T \in R^m$ $(m = 2,3)$ as the position of *nodei* . $P_i(i = 0,1,..., n-1)$ is the link radius of *nodei* (i.e., radio range of *nodei*). Thus, $\| x_i - x_j \|$ is defined as the Euclidean distance between *nodei* and *nodej* . Denote connection matrix as $A = (a_{ij})_{n\times n}$ which consists of the coefficients of interactions between the nodes. Here, the coefficients of interaction are taken as

$$a_{ij} = \begin{cases} 0, \| x_i - x_j \| > \min\{P_i, P_j\}, & \\ & (i \neq j), \\ 1, \| x_i - x_j \| \leq \min\{P_i, P_j\}, & \\ 0, (i = j). & \end{cases}$$

During positioning a WSN, the estimated connection matrix is changing and it is denoted as $\overline{A} = (\overline{a}_{ij})_{n\times n}$.

Definition 1 To locate a WSN with connection matrix $A = (a_{ij})_{n\times n}$, if the *rowi* of $\overline{A}$ is the same as A , we call the *nodei* is to meet the node-link structure of the WSN.

If a WSN is determined then the connection matrix $A = (a_{ij})_{n\times n}$ is also determined, i.e., the node-link structure of each node is determined. In addition, if $\overline{A} = A$ is satisfied, the localization is complete.

In practice, sensor nodes may be randomly scattered in special region, and most sensors do not know their locations. However, the connection matrix of WSNs can be easily obtained. In order to locate sensors, SSLA only needs some anchors and the information of connection matrix $A = (a_{ij})_{n\times n}$. Usually, three or more anchors are needed to locate a two-dimensional WSN or four or more anchors are needed to locate a three-dimensional WSN. In addition, there have some sleeping nodes which do not contact with each other in a WSN. The sleeping nodes and anchors cannot be moved during localization by SSLA.

From above mention, we can describe various cases for any *nodei* and *nodej* $(i \neq j)$ in the implementation of SSLA:

(1) If $a_{ij} = 1$, *nodei* and *nodej* are neighbours in the original WSN.

(2) If $a_{ij} = 0$, *nodei* and *nodej* are not neighbours in the original WSN.

During localization, we examine the estimated locations of *nodei* and *nodej* $(i, j = 0,1,\ldots,n-1)$.

If $a_{ij} = 1$, there are two cases shown in Figure 4.2 (a) and (c). If the case is as Figure 4.2 (a), *nodei* and *nodej* will not move. Otherwise, the algorithm will drive *nodei* and *nodej* to move towards each other, as shown in Figure 4.2 (c).

If $a_{ij} = 0$, there are two cases shown in Figure 4.2 (b) and (d). If the case is as Figure 4.2 (b), the algorithm will drive *nodei* and *nodej* to move against each other. Otherwise, *nodei* and *nodej* will not move, as shown in Figure 4.2 (d).

The various cases of movement of *nodei* and *nodej* during localization are shown in Figure 4.2.

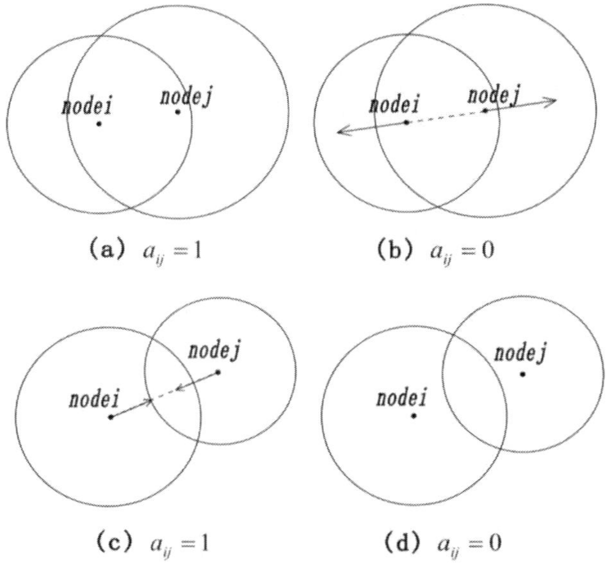

Figure 4.2. The various cases of movement of nodei and nodej ($i \neq j$).

From Figure 4.2, we get the iterative relations to describe the algorithm as follows.

If both *nodei* an *nodej* are anchors or sleeping nodes, we take

$$x_i(t_k) \leftarrow x_i(t_k), x_j(t_k) \leftarrow x_j(t_k);$$

otherwise, take

$$\begin{bmatrix} x_i(t_k) \\ x_j(t_k) \end{bmatrix} \leftarrow \begin{bmatrix} x_i(t_k) \\ x_j(t_k) \end{bmatrix} + \alpha(t_k)\beta(\alpha_{ij}, t_k)\begin{bmatrix} x_i(t_k) - x_j(t_k) \\ x_j(t_k) - x_i(t_k) \end{bmatrix},$$
$$(i, j = 0,1,\ldots,n-1, k \in Z). \qquad (*)$$

There is a double loop in the iterative relations. After the double loop, we take $k = k+1$.

In the iterative relation (*), we take

$$\beta(\alpha_{ij}, t_k) = \frac{1}{2}(2a_{ij} - 1 - \text{sgn}(\min\{P_i, P_j\} - \| x_i(t_k) - x_j(t_k) \|)) \quad ,$$
$$\beta(\alpha_{ij}, t_k) \in \{0,1,-1\}.$$

$\beta(\alpha_{ij}, t_k)$ determines the move directions of *nodei* and *nodej*, e.g., make *nodei* and *nodej* move towards each other as shown in Figure 4.2 (c) ($\beta(\alpha_{ij}, t_k) = 1$) and make *nodei* and *nodej* move against each other as shown in Figure 4.2 (b) ($\beta(\alpha_{ij}, t_k) = -1$), or determines the *nodei* and *nodej* to keep still as shown in Figs. 4.2 (a) and (d) ($\beta(\alpha_{ij}, t_k) = 0$).

$\alpha(t)$ is the control coefficient which plays the role of amplitude of each node which is described in Figure 4.3.

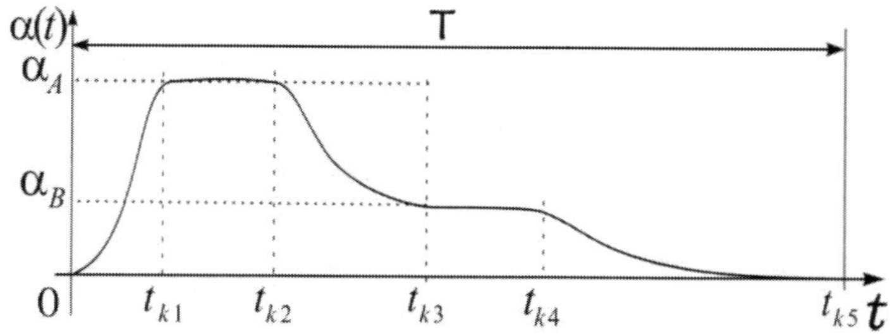

Figure 4.3. The curve of control coefficient $\alpha(t) \geq 0$.

The curve of control coefficient $\alpha(t)$ is called control curve, as shown in Figure 4.3. The control curve is divided into five stages. In stage $[0,t_{k1}]$, the control curve may quickly reach the maximum value α_A. Then, in stage $[t_{k1},t_{k2}]$, the control curve maintains a relatively long time without changes. In stage $[t_{k2},t_{k3}]$, the control curve goes down to value α_B. In stage $[t_{k3},t_{k4}]$, it maintains a relatively long time without changes. Lastly, the control curve slowly goes down to very small values close to 0 in stage $[t_{k4},t_{k5}]$. These five stages make up the localization process. The total time used is denoted as T. T can be taken in proportion to the number of nodes. There is no other demand for each stage. Here, we also use discrete values $\alpha(t_k)$ instead of continuous value (we take $\alpha_A = \frac{3}{2n}\sum_{i=0}^{n-1} P_i, \alpha_B = \frac{1}{2n}\sum_{i=0}^{n-1} P_i, \Delta t = t_k - t_{k-1} = \frac{T}{N}, t_0 = 0$ and N is a constant, e.g., we can take $N = 300$). Now, we give the flowchart of the main program of SSLA.

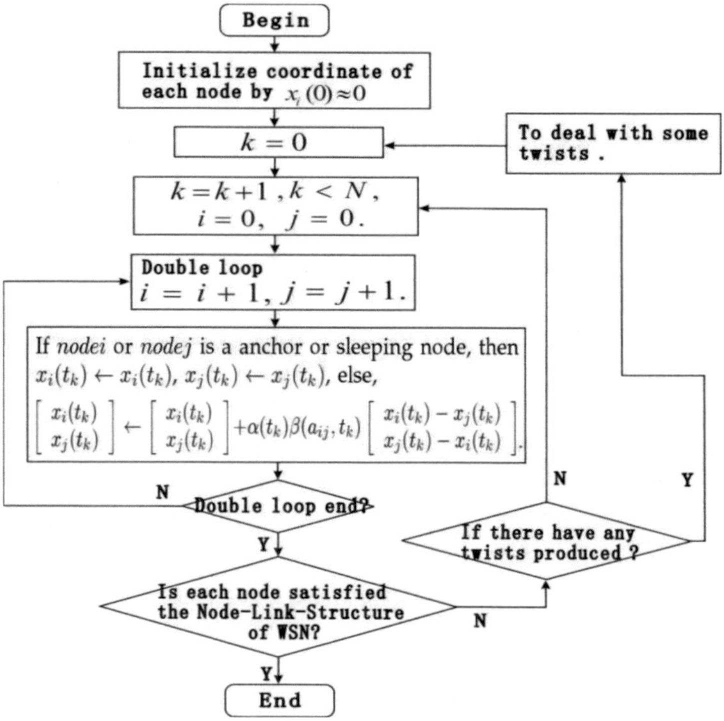

Figure 4.4. The flowchart of the main program of SSLA.

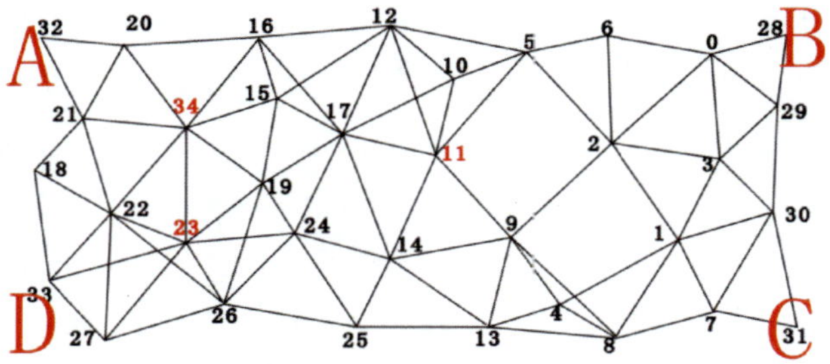

Figure 4.5. The original 35-node WSN in a rectangular region with each corner marked A, B, C, D, respectively.

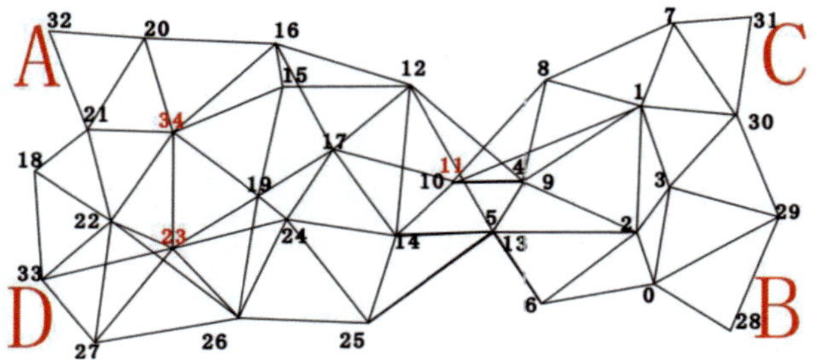

Figure 4.6. A twist appears during localization.

Note that in the SSLA program, initial locations of nodes are initialized as $x_i(0) \approx 0$ $(i = 0,1,\ldots,n-1)$. However, for any other initial locations of nodes, the results of calculation will be not affected. To some extent, this process of SSLA converges for any initial locations.

Here, we explain how to deal with twists in Figure 4.4 and why to take control curve as Figure 4.3. As for twists and its treatment, Figs. 4.5 and 4.6 show the details.

During localization of the 35-node WSN as shown in Figure 4.5 ($node11, node23, node34$ are anchors), the phenomenon of regional folding may happen as in Figure 4.6, which we call a twist. A twist is just like an unstable equilibrium, as long as some data is modified in the corresponding region, nodes will soon find their own stable equilibrium position and twists will disappear. If we do not to deal with twists, then localization will fail. It is easy to judge whether a twist appear and to solve a twist by the SSLA. In addition, if the network is strong enough, twists may not appear. In general, if twists appear during localization, there may be only a few or only one. The SSLA program supports to modify the coordinates of some nodes to solve the problem of twists. For example, to deal with the twist like Figure 4.6, the SSLA program can modify the regional data of B (or C), and then the problem of twists will be solved.

Here, we give some illustrations for control curve in Figure 4.3. In the SSLA program, all nodes are forced to separate in stage $[0, t_{k1}]$, then each node oscillates around its equilibrium position in stage $[t_{k1}, t_{k2}]$. Next, the oscillation amplitude of each node gradually decreases in stage $[t_{k2}, t_{k3}]$ and each node continuously oscillates with a relatively small amplitude in stage $[t_{k3}, t_{k4}]$. Lastly, the amplitude of each node would approach to zero and the WSN will be located in stage $[t_{k4}, t_{k5}]$.

The algorithm drives node complex interactions in a WSN. For nodes to escape from the restrictions of other nodes to find their own equilibrium position, we take the maximum amplitude bigger than the average radio range, i.e. $\alpha_A = \frac{3}{2n}\sum_{i=0}^{n-1} P_i > \frac{1}{n}\sum_{i=0}^{n-1} P_i$. It is a key value of α_A in the control curve. A large number of numerical simulations show that the value is appropriate.

In fact, we can take various shapes of the control curve, but the control curve must tend to 0 as time increases. However, a large number of experimental results show that the shape of control curve in Figure 4.3 is more effective.

Here, denote $\bar{x}_i$ as the calculation results of x_i by SSLA and some definitions are given as follows.

Definition 2 $\Delta(x) = (\Delta(x_{i1}), \ldots, \Delta(x_{im})) = (\frac{1}{n}\sum_{i=0}^{n-1}(x_{i1} - \bar{x}_{i1}), \ldots, \frac{1}{n}\sum_{i=0}^{n-1}(x_{im} - \bar{x}_{im}))^T$ is called the total average relative error.

Definition 3 $\delta(x) = (\delta(x_{i1}), \ldots, \delta(x_{im})) = (\frac{1}{n}\sum_{i=0}^{n-1}|x_{i1} - \bar{x}_{i1}|, \ldots, \frac{1}{n}\sum_{i=0}^{n-1}|x_{im} - \bar{x}_{im}|)^T$ is called the total average absolute error.

Definition 4 $\sigma = \sigma(x) = \frac{1}{n}\sum_{i=0}^{n-1}\sqrt{(x_{i1} - \bar{x}_{i1})^2 + \ldots + (x_{im} - \bar{x}_{im})^2}$ is called the total average distance error.

Definition 5 $\rho = \rho(x) = \dfrac{n \times \sigma(x)}{\sum_{i=0}^{n-1} P_i} \times 100\,\%$ is called the positioning accuracy.

It is obvious that $\Delta(x), \delta(x), \sigma(x)$ and $\rho(x)$ can reflect different levels of positioning accuracy. Especially, the smaller of the values $\sigma(x)$ and $\rho(x)$, the better the positioning accuracy.

For two-dimensional distribution WSNs, the (x_i, y_i) denote the coordinates of *nodei* $(i = 0,1,\ldots,n-1)$ and we initialize $(x_i(0), y_i(0)) \approx (0,0)$. In this paper, if there are no special instructions, nodes are randomly distributed in WSNs.In the following, we give some examples to analyze the algorithm of SSLA. We locate a 100-node WSN with nodes randomly distributed in a 400 ×200 area and the radio range of each node is 80. Here, we take 6 anchors $(node18, node33, node62, node91, node93, node99)$ and use the SSLA to locate the 100-node WSN. The positioning process is shown in Figures. 4.7 – 4.9.

The calculation results by SSLA are shown as follows.

Total average relative error:

$\Delta(x) = -0.89898$, $\Delta(y) = 1.65683$.

Total average absolute error:

$\delta(x) = 2.78845$, $\delta(y) = 3.31648$.

Total average distance error:

$\sigma = 5.04485$.

Average positioning accuracy:

$\rho = 5.04485/80 = 6.3060625\%$.

The localization for this 100-node WSN by SSLA is successful. The localization process tracks of coordinates $x_i, y_i (i = 0,..,99)$ are shown in Figure 4.10 and 4.11. Figures. 4.10 and 4.11 show the relation between the 5 stages of control curve $\alpha(t)$ and the localization process tracks of coordinates

$x_i, y_i (i = 0,..,99)$. In the following, we analyze the localization algorithm of SSLA.The positioning accuracy of SSLA is related to the connectivity strength (i.e., the node density or average node degree) of network and the ratio of the anchors. For randomly created WSNs with similar network connectivity strength and anchor conditions, SSLA has little effect on the positioning accuracy. We investigate the effect of radio range on the positioning accuracy. To locate a WSN with 250 sensor nodes randomly distributed in a 1000×1000 square area, we take 10 anchors randomly.

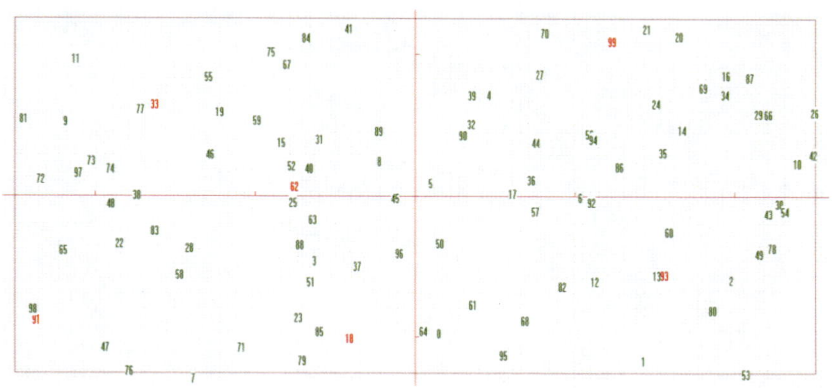

Figure 4.7. 100 nodes randomly distributed in a 400×200 area, each with a radio range of 80.

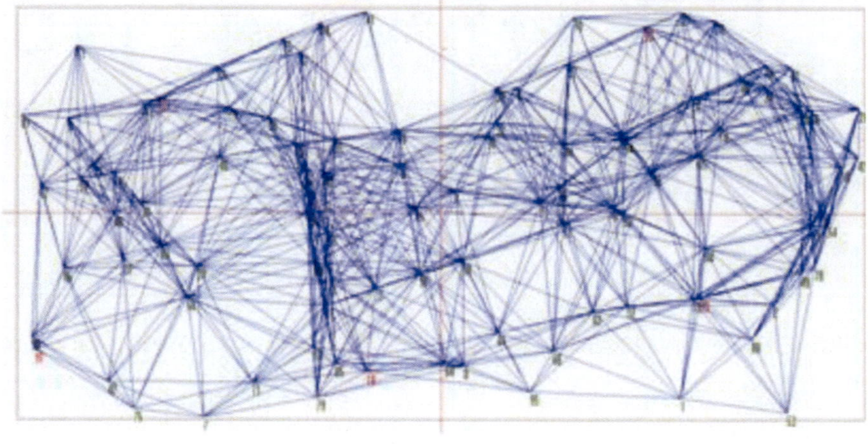

Figure 4.8. The original link network of 100-node WSN in the 400×200 area.

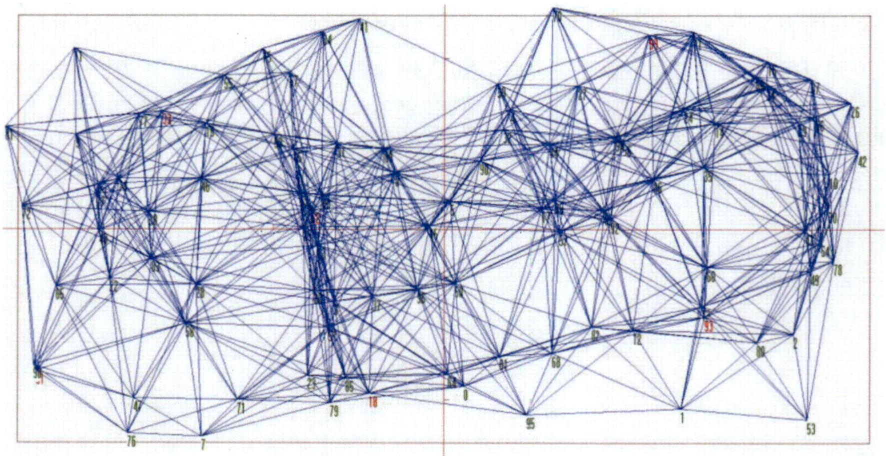

Figure 4.9. Localization result of the 100-node WSN by SSLA.

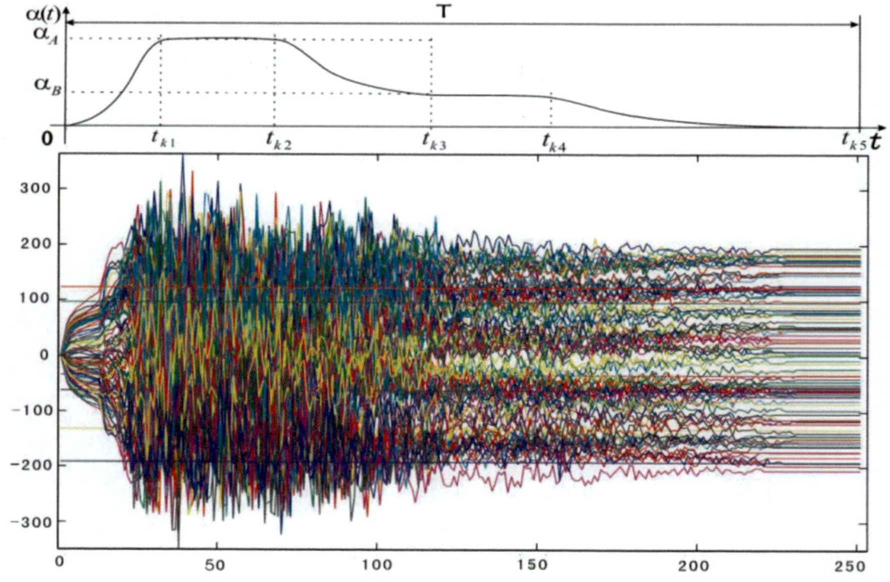

Figure 4.10. The localization track curves of coordinates $x_i(i = 0,..,99)$ and the control curve $\alpha(t)$.

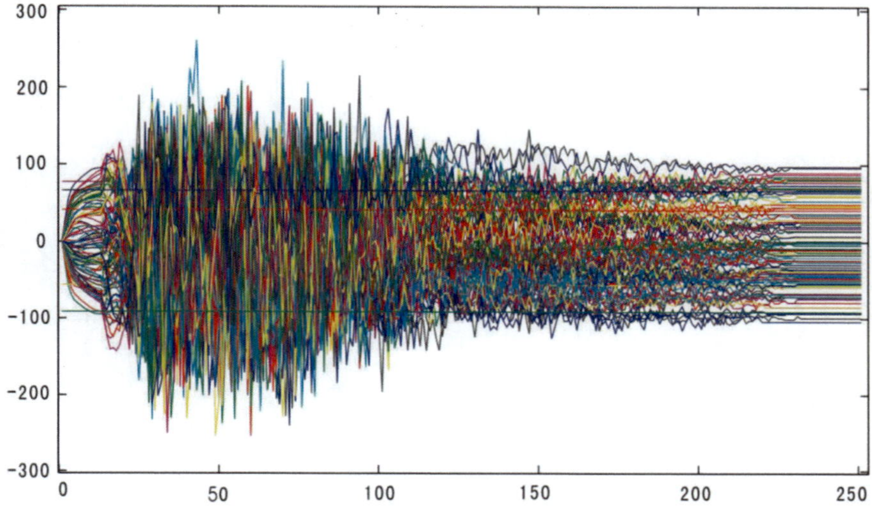

Figure 4.11. The localization track curves of coordinates $y_i (i = 0,...,99)$.

In addition, we assume that the radio range of each node is r. We take various values of radio range, ie, $r = 100 + k \times \Delta$ $(\Delta = 10,\quad k = 0,1,...10)$, respectively.

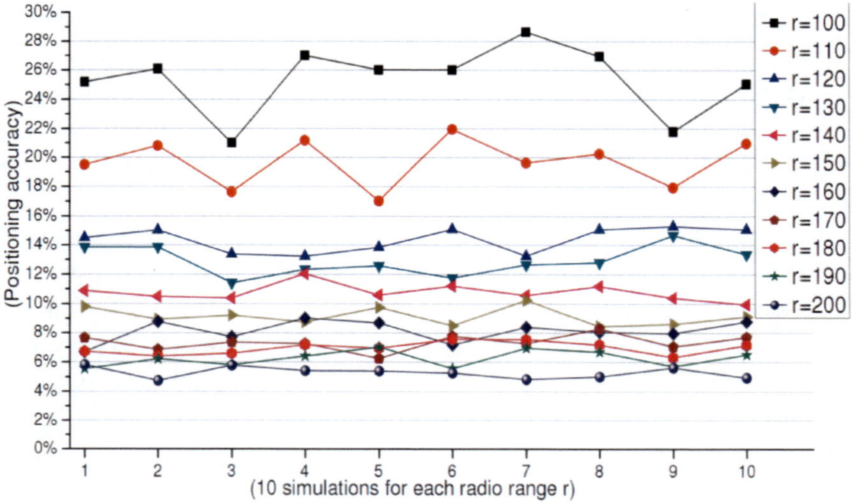

Figure 4.12. Investigation the effect of radio ranges on positioning accuracy for 250-node WSNs (take 10 anchors randomly).

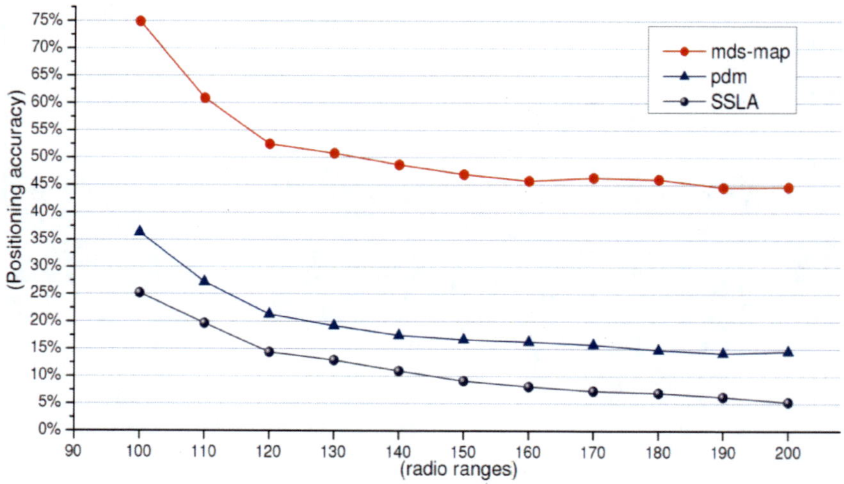

Figure 4.13. The positioning accuracy vs. radio range.

For each radio range r, we randomly create 10 WSNs. The results calculated by SSLA are shown in Figure 4.12.

When the radio range $r = 100$, the positioning accuracy is about $(24 \pm 5)\%$. When the radio range $r = 200$, the positioning accuracy is about $(5 \pm 1)\%$. Figure 4.12 shows the positioning accuracy calculated by the SSLA is more stable for different radio ranges. With the network connectivity strength increasing, the fluctuations of positioning accuracy and positioning accuracy become smaller.

We calculate the average positioning accuracy of 10 simulations and compare SSLA with MDS-MAP and PDM. The results are shown in Figure 4.13.

It is obvious that SSLA can obtain a higher positioning accuracy for locating of a WSN. In addition, we investigate the relation between the positioning accuracy and different ratio of anchors. We take WSNs with 250 sensor nodes randomly distributed in a 1000×1000 square area. We consider two cases of radio ranges: (a) $r = 100$; (b) $r = 130$. Each of the networks is randomly created and anchors are randomly distributed in a specific WSN. The relation of positioning accuracy and different ratio of anchors is shown in Figure 4.14.

From Figure 4.14, when the ratio of anchors is more than 2%, the positioning accuracy is relatively stable. The stronger connectivity of a WSN is, the higher positioning accuracy can be obtained. To locate WSNs, SSLA only needs a low ratio of anchors.

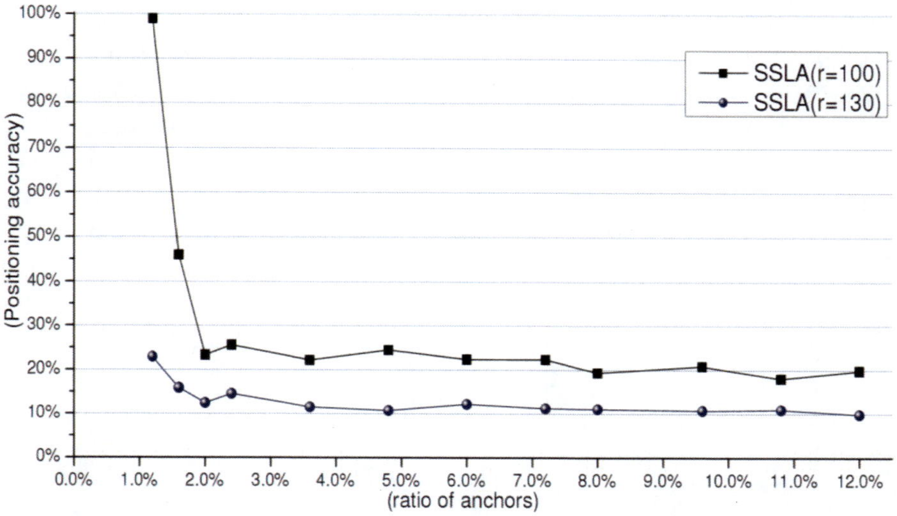

Figure 4.14. The positioning accuracy vs. ratio of anchors.

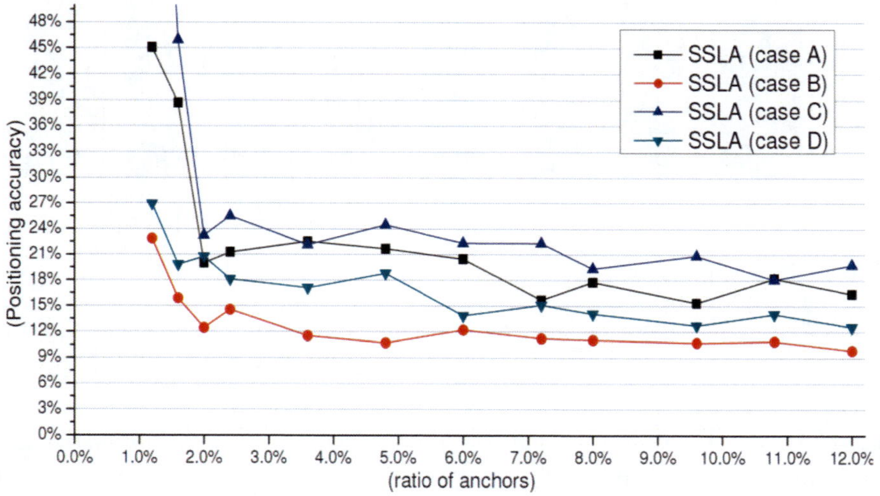

Figure 4.15. Four cases of localization of WSNs using SSLA. Case A: Different radio ranges of each node: radio ranges randomly form 100 to 130; case B: The same radio range of each node: r=130; case C: The same radio range of each node: r=100; case D: In the left half of square area, radio range of each node r=100, in the right half of square area, radio range of each node r=130.

Next, we investigate the algorithm for positioning anisotropic WSNs. We also take WSNs with 250 sensor nodes randomly distributed within a 1000×1000 square area. We study four cases of the WSNs, and the results are shown in Figure 4.15.

Figure 4.15 shows that the algorithm also can achieve stable, high positioning accuracy in all kinds of networks even with a low ratio of anchors.

Lastly, we investigate the algorithm for large-scale networks localization. We take three cases of networks: (a) Ratio of anchors 5%, average node degree 8; (b) Ratio of anchors 10%, average node degree 10; (c) Ratio of anchors 10%, average node degree 12. We take number of nodes from 100 to 5000 for each case. The results are shown in Figure 4.16.

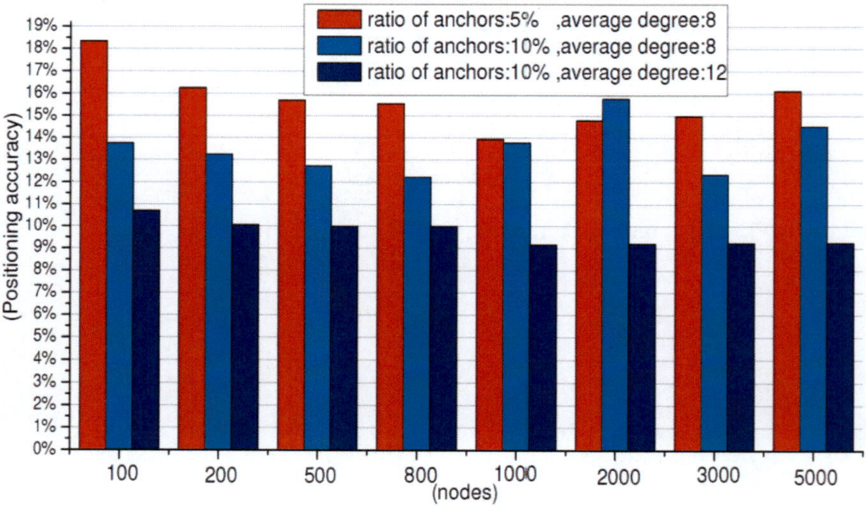

Figure 4.16. The positioning accuracy vs. network.

We draw a conclusion that the algorithm positioning sensor networks is related to the connectivity strength of networks, and the positioning error of SSLA will not accumulate as the number of nodes increases. SSLA is a stable algorithm which requires only a low ratio of anchors and it is suitable for positioning large-scale networks.

To solve the problem of network localization, there are several key questions raised in ref. [78] and it points out that answers to any of these questions would be an important step toward producing practical localization algorithms.One of question is: Is there an efficient algorithm for approximate localization in spare graphs, either by permitting moderate errors on distances or by permitting the algorithm to misplace some small fraction of the sensors? In this paper, we give the answer: Yes. Moreover, there is no any fraction of the sensors that will be misplaced by the algorithm.

SSLA is an effective rang-free algorithm. It belongs to a centralized algorithm. To locate a WSN, we just need a small number of anchors. Of course, more anchors, as well as stronger connection of WSNs will increase the positioning accuracy.

The algorithm needs relatively small computational cost. The computational complexity is $O(n^2)$. Our work shows that the positioning accuracy of SSLA can reach 10% or even higher. When the reference nodes are moved, SSLA does not have to reposition the entire network, and only needs to modify the corresponding local data.

In addition, we believe that SSLA also has wide applications in other areas, such as in a fixed structure formation and in the field of biological molecular structure. In particularly, according to the known network topology and some small portion of sensor locations, to find the locations of the other sensors has great meaningful in the field of aerospace. In the future, we will further discuss and study the applications of the algorithm in other fields.

CONCLUSION

In the past few years, advances in complex networks have uncovered some amazing similarities among diverse complex networks such as the Internet, cellular networks, metabolic networks, protein-protein interaction network, and even the community of Hollywood movie stars. To uncover the generic properties of different kinds of complex networks, many quantities and measures have been proposed to characterize the structural properties of networks in recent years.

However, measuring these basic properties of a complex network is just the first step toward understanding its structure. The next step, then, is to develop an evolution model with a topology of similar statistical properties, thereby obtaining a platform on which related problem can be studied. Unfortunately, it is usually lack of clear quantitative relationships in the complex network modeling, but only some descriptive terms are given therein. This results in the construction method of networks should be different from that used to build up the mathematical models in the past based on the quantitative relationships in various scientific fields, such as physics and mechanics. In this chapter, we have provided a new method to construct network models via simple rules.

The main idea of the approach is that some simple rules are firstly extracted from the descriptive terms, and then network models are constructed via these simple rules. Two examples have been given to illustrate the research approach.

One example is about the modeling of protein-protein interaction networks and the other example is about the positioning of wireless sensor networks.

Our results have shown that our research approach is feasible. Finding some simple rules from descriptive terms to construct network models may become a new research direction of complex networks modeling. Moreover, this direction may bring about a new way of thinking for scientific research.

ACKNOWLEDGMENTS

This work was supported by the National Science Foundation of China (Grant nos. 10832006, 11172158). ZRL conceived and designed the research, SMC, QBH and ZRL wrote the paper, SMC and QBH contributed equally to this work.

REFERENCES

[1] Strogatz, S.H. (2001). Exploring complex networks. *Nature*, 410, 268-276.

[2] Albert, R., and Barabasi, A.L. (2002). Statistical mechanics of complex networks. *Rev. Mod. Phys.*, 74, 47-91.

[3] Newman, M.E.J. (2003). The structure and function of complex networks. *SIAM Rev.*, 45, 167-256.

[4] Erdos, P., and Renyi, A. (1959). On the evolution of random graphs. *Publ. Math. Inst. Hung. Acad. Sci.*, 5, 17-60.

[5] Wang, X., and Chen, G. (2003). Complex networks: small-world, scale-free and beyond. *IEEE Circ. Syst. Mag.*, 3, 6-20.

[6] Costa, L.F., Rodrigues F.A., andTravieso G., *et al.* (2007). Characterization of complex network: a survey of measurements. *Adv. Phys.*, 56, 167-242.

[7] Boccaletti, S., Latora, V., and Moreno, Y., *et al.* (2006). Complex networks: Structure and dynamics. *Phys. Rep.*, 424,175-308.

[8] Watts D.J., and Strogatz S.H. (1998). Collective dynamics of 'small world' networks. *Nature*, 393, 440-442.

[9] Barabasi, A.L., and Albert R. (1999). Emergence of scaling in random networks. *Science*, 286, 509-512.

[10] Ravasz, E., Somera A.L., and Mongru, D.A., *et al.* (2002). Hierarchical organization of modularity in metabolic networks. *Science*, 297, 1551-1555.

[11] Ravasz, E., and Barabasi, A.L. (2003). Hierarchical organization in complex networks. *Phys. Rev. E*, 67, 026112.

[12] Maslov, S., and Sneppen K. (2002). Specificity and stability in topology of protein networks. *Science*, 296, 910-913.

[13] Wan, X., Cai, S., and Zhou, J., et al. (2010). Emergence of modularity and disassortativity in protein-protein interaction networks. *Chaos*, 20, 045113.

[14] He, Q., Chen, F., and Cai, S., *et al.* (2011). An efficient range-free localization algorithm for wireless sensor networks. *Sci. China Tech. Sci.,* 54, 1053-1060.

[15] Bollobas, B. (1998). Modern Graph Theory. Springer: New York.

[16] Wasserman, S., and Faust, K. (1994). *Social Networks Analysis*. Cambridge University Press, Cambridge.

[17] Pastor-Satorras, R., Vazquez, A., and Vespignani, A. (2001). Dynamical and correlation properties of the internet. *Phys. Rev. Lett.*, 87, 258701.

[18] Newman, M.E.J. (2003). Mixing patterns in networks. *Phys. Rv. E*, 67, 026126.

[19] Prachumwat, A., and Li, W.H. (2006). Protein function, connectivity, duplicability in Yeast. *Mol. Biol. Evol.*, 23, 30-39.

[20] Barabasi, A.L., and Oltvai, Z.N. (2004). Network biology: understanding the cell's functional organization. *Nat. Rev. Genet.,* 5, 101-113.

[21] Yook, S.H., Oltvai, Z.N., and Barabasi, A.L. (2004). Functional and topological characterization of protein interaction networks. *Proteomics*, 4, 928-942.

[22] Uetz, P., Giot, L., and Cagney, G., *et al.* (2000). A comprehensive analysis of protein-protein interactions in *Saccharomyces cerevisiae*. *Nature*, 403, 623-627.

[23] Ito, T., Chiba, T., and Ozawa R., *et al.* (2001). A comprehensive two-hybrid analysis to explore the yeast protein interactome. *Proc. Natl. Acad. Sci. U.S.A.*, 98, 4569-4574.

[24] Guldener, U., Munsterkotter, M., and Oesterheld, M., *et al.* (2006). MPact: the MIPS protein interaction resource on yeast. *Nucleic Acids Res.*, *34*, D436-441.

[25] Li, S., Armstrong, C.M., and Bertin, N. *et al.* (2004). A map of interactome network of the metazoan *C. elegans*. *Science*, 303, 540-543.

[26] Giot, L., Bader, J.S., and Brouwer, C. et al. (2003). A protein interaction map of drosophila melanogaster. *Science*, 302, 1727-1736.

[27] Stelzl, U., Worm, U., and Lalowski, M. *et al.* (2005). A human protein-protein interaction network: a resource for annotating the proteome. *Cell*, 122, 957-968.

[28] Rual, J.F., Venkatesan, K., and Hao, T. *et al.* (2005). Towards a proteome-scale map of the human protein-protein interaction network. *Nature*, 437, 1173-1178.

[29] Wagner, A. (2001). The yeast protein interaction network evolves rapidly and contains few redundant duplication genes. *Mol. Biol. Evol.*, 18, 1283-1292.

[30] Jeong, H., Mason, S.P., and Barabasi, A.L., *et al.* (2001). Lethality and centrality in protein networks. *Nature*, 411, 41-42.

[31] Noort, V., Snel, B., and Huynen, M.A. (2004). The yeast coexpression network has a small-world, scale-free architecture and can be explained by a simple model. *EMBO reports*, 5, 1-5.

[32] Hase, T., Niimura, Y., and Kaminuma, T., *et al.* (2008). Non-uniform survival rate of heterodimerization links in the evolution of the yeast protein-protein interaction network. *PLoS ONE*, 3, e1667.

[33] Ohno, S. (1970). *Evolution by gene duplication.* Springer: Berlin.

[34] Teichmann, S.A., and Babu, M.M. (2004). Gene regulatory network growth by duplication. *Nature genet.*, 36, 492-496.

[35] Brenner, S.E., Hubbard, T., and Murzin, A., *et al.* (1995). Gene duplications in *H.* influenzae. *Nature,* 378, 140.

[36] Teichmann, S.A., Park, J., and Chothia, C. (1998). Structural assignments to the *Mycoplasma genitalium* proteins show extensive gene duplications and domain rearrangements. *Proc. Natl. Acad. Sci. USA*, 95, 14658-14663.

[37] Gough, J., Karplus, K., and Hughey, R., et al. (2001). Assignment of homology to genome sequences using a library of hidden Markov models that represent all proteins of known structure. *J. Mol. Biol.*, 313, 903-919.

[38] Sole, R.V., Pastor-Satorras, R., and Smith, E., *et al.* (2002). A model of large-scale proteome evolution. *Adv. Complex Syst.*, *5*, 43-54.

[39] Kim, J., Krapivsky, P.L., and Kahng, B., *et al.* (2002). Infinite-order percolation and giant fluctuations in a protein interaction network. *Phys. Rev. E*, 66, 055101 (R).

[40] Vazquez, A., Flammini, A., and Maritan, A., *et al.* (2003). Modeling of protein interaction networks. *ComPlexUs*, 1, 38-44.

[41] Chung, F., Lu L., and Dewey, T.G., *et al.* (2003). Duplication models for biological networks. *J. Comput. Biol.*, 10, 677-687.

[42] Pastor-Satorras, R., Smith, E., and Sole, R.V. (2003). Evolving protein interaction networks through gene duplication. *J. Theor. Biol.*, 222, 199-210.

[43] Colizza, V., Flammini, A., and Maritan, A. (2005). Characterization and modeling of protein-protein interaction networks. *Physica A*, 352, 1-27.

[44] Bebek, G., Berenbrink, P., and Cooper, C., *et al.* (2006). The degree distribution of the generalized duplication model. *Theor. Comput. Sci., 369*, 234-249. [44]. Ispolatov, I., Krapivsky, P.L., and Yuryev, A. (2005). Duplication-divergence model of protein interaction network. *Phys. Rev. E*, 71, 061911.

[45] Ispolatov, I., Krapivsky, P.L., and Mazo, I., et al. (2005). Cliques and duplication-divergence network growth. *New J. Phys.*, 7, 145.

[46] Takemoto, K., and Oosawa, Chikoo. (2005). Evolution networks by merging cliques. *Phys. Rev.* E, 72, 046116.

[47] Takemoto, K., and Oosawa, Chikoo. (2007). Modeling for evolving biological networks with scale-free connectivity, hierarchical modularity, and disassortativity. *Math. Biosci.*, 208, 454-468.

[48] Evlampiev, K., and Isambert, H. (2008). Conservation and topology of protein interaction networks under duplication-divergence evolution. *Proc. Natl. Acad. Sci. USA*, 105, 9863-9868.

[49] Xu, C., Liu, Z., and Wang, R. (2010). How divergence mechanisms influence disassortative mixing property in biology. *Physica A*, 389, 643-650.

[50] Kellis, M., Patterson, N., and Endrizzi, M., *et al.* (2003). Sequencing and comparison of yeast species to identify genes and regulatory elements. *Nature*, 423, 241-254.

[51] Force, A., Lynch, M., and Pickett, F.B., *et al.* (1999). Preservation of duplicate genes by complementary, degenerative mutations. *Genetics*, 151, 1531-1545.

[52] Cai, S. (2012). *PhD thesis*, Shanghai University, Shanghai.

[53] Capkun, S., Hamdi, M., and Hubaux, J.P.(2002). GPS-Free positioning in mobile ad-hoc networks. *Cluster Computing*. 5(2),157-167.

[54] Girod, L.,and Estrin, D.(2001). Robust range estimation using acoustic and multimodal sensing. Proc. of the IEEE International Conference on Intelligent Robots and Systems, Maui, Hawaii, USA, 1312-1320.

[55] Priantha, N.B., Chakraborty, A., and Balakrishnan, H.(2000). The cricket location-support system. Proc. 6th ACM International Conference on Mobile Computing and Networking (ACM MOBICOM), Boston, MA, 32-43.

[56] Savvides, A., Han, C.C.,and Srivastava, M.B.(2001). Dynamic finge grained localization in ad-hoc networks of sensors. Proc. 7th Annual Int'1 Conf on Mobile computing and Networking(MoviCom),Rome, Italy, 166-179.

[57] Savvides, A., Park, H., Mani, B.,and Srivastava, M.(2002). The bits and flops of the N-hop multilateration primitive for node localization problems. Proc. 1st ACM Int'1 Workshop on Wireless Sensor Networks and Application,Atlanta,GA, 112-121.

[58] Niculescu, D.,and Nath, B.(2003). Ad hoc positioning system (APS) using AOA. *Proc. 22nd* Annual Joint Conf. the IEEE Computer and Communications Societies (IEEE INFOCOM 2003), 1744-1753.

[59] Priyantha, N.B., Miu, A.K.L., Balakrishnan, H., et al.(2001). The cricket compass for context-aware mobile applications. Proc. of the 7th Annual Int'l Conf. on Mobile Computing and Networking, Rome: ACM Press, 1-14.

[60] Bahl, P.,and Padmanabhan, V.N.(2000). RADAR: An in-building RF-based user location and tracking system. Proc. of IEEE Infocom 2000, Tel Aviv, Israel, 775-784.

[61] Girod, L., Bychovskiy, V., Elson, J., et al.(2002). Locating tiny sensors in time and space: A case study. Proc. of the 2002 IEEE Int'l Conf. on Computer Design: VLSI in Computers and Processors. Freiburg: IEEE Computer Society, 214-219.

[62] Galstyan, A., Krishnamachari, B., Lerman, K., et al.(2004). Distributed online localization in sensor networks using a moving target. Proc. Third Internat. Symp. on Information Processing in Sensor Networks, ACM Press, New York, 61-70.

[63] Nagpal, R., Shrobe, H.,and Bachrach, J.(2003). Organizing a global coordinate system from local information on an ad hoc sensor network. Proc. in Sensor Networks: Second International Workshop, IPSN 2003, No. 2634, Lecture Notes in Computer Science, Springer, Palo Alto, 333-348.

[64] Meesookho, C., Mitra, U.,and Narayanan, S.(2008). On energy-based acoustic source localization for sensor network. *IEEE Transactions on Signal Processing*, 56(1),365-377.

[65] Langendoen, K.,and Reijers, N.(2003). Distributed localization in wireless sensor networks: a quantitative comparison. *Computer Networks*, 43(4),499-518.

[66] Sheng, X.H.,and Hu, Y.H.(2005). Maximum likelihood multiple-source localization using acoustic energy measurements with wireless sensor networks. *IEEE Transactions on Signal Processing*,53(1),44-53.

[67] Bulusu, N., Heidemann, J.,and Estrin, D.(2000). GPS-less low cost outdoor localization for very small devices. *IEEE Personal Communications Magazine*,7,28-34.

[68] Doherty, L.(2000). Algorithms for position and data recovery in wireless sensor networks. Masters Report, University of California, Berkeley, CA.

[69] Nicolescu, D.,and Nath, B.(2001). Ad-hoc positioning systems (APS). Proc. of the 2001 IEEE Global Telecommunications Conf. San Antonio: IEEE Communications Society, 5,2926-2931.

[70] Niculescu, D.,and Nath, B.(2003). DV based positioning in ad hoc networks. *Journal of Telecommunication Systems*, 22,267-280.

[71] He, T., Huang, C., Blum, B.M., et al.(2003). Range-free localization schemes for large scale sensor networks. Proc. 9th Annual International Conference on Mobile Computing and Networking, 81-95.

[72] Shang, Y., Ruml, W., Zhang, Y., et al.(2003). Localization from mere connectivity. Proc. of the Fourth ACM Internat. Symp. on Mobile ad hoc Networking and Computing MOBIHOC 2003, ACM Press, NewYork, 201-212.

[73] Lim, H.,and Hou, J.(2009). Distributed localization for anisotropic sensor networks. *ACM Transactions on Sensor Networks*, 5(2),1-26.

[74] Stupp, G.,and Sidi, M.(2005). The expected uncertainty of range-free localization protocols in sensor networks. *Theoretical Computer Science*, 344(1),86-99.

[75] Doherty, L., Pister, K.S.J.,and Ghaoui, L.E.(2001), Convex position estimation in wireless sensor networks. Proc. of the IEEE INFOCOM 2001. Anchorage: IEEE Computer and Communications Societies,1655-1663.

[76] Eren, T., Goldenberg, D., Whiteley, W., et al.(2006). Rigidity and randomness in network localization. *IEEE INFOCOM*, 4:2673-2684.

[77] Anderson, B.D.O., Belhumeur, P.N., Eren, T., et al.(2009). Graphical properties of easily localizable sensor networks. *Wireless Networks*, 15(2):177-191.

[78] Aspnes, J., Eren, T., Goldenberg, D.K., et al.(2006). A theory of network localization. *IEEE Transactions on Mobile Computing*, 5(12),1663-1678.

[79] Vincent, T., Cheng, K.Y.,and Lui, K.S.(2006). Using micro-genetic algorithms to improve localization in wireless sensor networks. *Journal of Communications,* 1(4),1-10.

[80] Nan, G.F., Li, M.Q.,and Li, J.(2007). Estimation of node localization with a real-coded genetic algorithm in WSNs. Proc. the Sixth International Conference on Machine Learning and Cybernetics, Hong Kong, 19-22 August 2007,873-878.

[81] Yun, S., Lee, J., Chung, W., et al.(2009). A soft computing approach to localization in wireless sensor networks. Expert Systems with Applications, 36(4),7552-7561.

[82] Chuang, P.J.,and Wu, C.P.(2008). An effective PSO-based node localization scheme for wireless sensor networks. Proc. 2008 Ninth International Conference on Parallel and Distributed Computing, Applications and Technologies,187-194.

[83] Frand, H.H.(1974). Avian Flight Formati, *Bird-Banding,* 45(2), 160-169.

[84] Götmark, F., Winkler, D.W.,and Andersson, M.(1986). Flock-feeding on fish schools increases individual success in gulls, *Nature,* 319, 589-591.

[85] Olfati-Saber, R., Fax, J.A.,and Murray, M.M.(2007). Consensus and Cooperation in Networked Multi-Agent Systems, *Proceedings of the IEEE*, 95(1),215-233.

In: Network Topologies
Editor: Casandra Rendell

ISBN: 978-1-62618-180-9

Chapter 2

THE APPLICATION OF ARTIFICIAL NEURAL NETWORKS IN THE DIAGNOSIS OF CORONARY HEART DISEASE

S. G. Gorokhova[1*], A. G. Sboev[2], K. A. Kukin[2], R. B. Rybka[2], E. V. Muraseeva[1] and O. Yu. Atkov[3]

[1]I.M. Sechenov First Moscow State Medical University
[2]MEPhI National Research Nuclear University
[3]N.I. Pirogov Russian State Medical University

ABSTRACT

The present research is aimed to develop an ANN diagnostic model for the coronary atherosclerosis and ischemia for patients after coronary angiography on the basis of genetic, clinical laboratory and instrumental examination data. The analysis of the correlation between the signs allowed us to choose the factors most closely connected with the diagnosis. Hierarchical clustering and correlation analysis were adapted to allocate typical fields of diagnostic factors. Various types of ANN topologies (MLP, SVM, PCA, and hybrid network) were analyzed; we have found that the models based on ANN with principal components analysis, and double-layer perceptron ANN optimized with genetic algorithms achieve the best diagnostic efficacy.

* Corresponding author; phone: phone.: +7 903 597-91-95, e-mail: cafedra2004@mail.ru.

INTRODUCTION

During the last decades, artificial neural networks (ANNs) have become wildly used for identification classification, diagnosis and prediction purposes in various branches of medicine, i.e. radiology, oncology, cardiology, urology, gastroenterology etc. [1-11]. Cardiology has a special place in this list. In spite of the recent reducing trend, cardiovascular diseases, and especially coronary heart disease (CHD), remain the most prevalent cause of morbidity, disability, and mortality among persons in active working age all over the world [12-16]. In this regard, World Health Organization describes the development of innovative approaches to the prevention, diagnosis and treatment of CHD as one of the most important fields of work, and many efforts are pooled together to find effective ways to solve these problems. However, there is an evident and continuous growth of costs of medical care, therefore cost-effective and equitable health care innovations are preferred; artificial neural networks are one of them.

In health care practice, the more errors are made during decision making, the higher are the expenses. Wrong medical conclusions and inaccurate strategy of treatment may lead into significant financial losses, not to mention losses of human lives. Unfortunately, such errors are inevitable[1], because there is always a grain of uncertainty, as well as a probability, that the applied diagnostic techniques are insufficiently precise. The problem cannot be solved by the increase in number of such tests, because the uncertainty grows with the amount of information. Besides, a physician may be insufficiently experienced or do not have enough practice in analyzing clinical information and drawing conclusions. An experienced doctor can also be misled. Even a confident grasp of differential diagnostic search does not eliminate a possibility of mistake, because of occasional clinical events and contradictions between facts and hypothesis that cannot be checked and confirmed at time. There is a high probability of mistakes under the conditions of extreme lack of time or excessive stress.

Despite the recent breakthrough in the diagnostic techniques, precise CHD diagnostics is still a serious problem in cardiology [18-22]. TOPIC study showed that 17.5% of diagnostic hypothesis formulated within the first minutes of the initial encounter were modified after the one year follow-up [23]. Our recent analysis was performed in the department, in which patients

[1] A rate of diagnostic error that ranges from 5% in the pathology, radiology and dermatology, and up to 10% to 15% in most other fields [17].

with cardiovascular pathology were hospitalized for the expert medical examination. It showed that the CHD was confirmed only in 18.5% of patients. In all other cases, in-depth examination by Holter monitoring, exercise and medicated stress tests (ECG, echocardiography, scintigraphy) and coronary arteriography in some cases, did not reveal any data supporting myocardial ischemia; cardialgia could be caused by some other heart diseases or extracardiac events.

In the light of the above, a great attention is paid to the ANN application to solve those diagnostic problems, in which the techniques of mathematical analysis can be applied. ANNs can render a substantial support for the physician, because they enable more complete and precise consideration of selected events and factors, reveal the relationships and dependencies between them in various states of the analyzed object, build flexible models, objectify the decision in this field. Among the directions of ANN application are biomedical signal processing, diagnosis of diseases, prognosis, and aiding medical decision support systems [24-26].

A good classical example is a prospective, blinded research by WG Baxt, which enrolled adult patients, hospitalized in an emergency department with anterior chest pain [27]. The diagnosis of acute myocardial infarction made by ANN and physicians were compared. It revealed that the proposed ANN model precisely recognizes myocardial infarction and excels the human in this respect. The physicians had a diagnostic sensitivity of 77.7% and a diagnostic specificity of 84.7%, while the artificial neural network 97.2% and 96.2%, respectively.

There are many causes that make diagnosis of CHD quite a difficult task, and first of them is the multiplicity of input data the physician should work with. A single case of clinical assessment of anginal pain includes the estimation of its character, localization, duration, conditions of its appearance and arrest [29].

Each of these features has several variants, which define if the pain results from the typical or atypical angina, or extracardiac cause. The typical angina is accompanied by attacks of pain, pressing retrosternal pains or discomfort and physical or emotional stress, which lasts for several minutes to a half an hour, and quickly passes away after nitroglycerine treatment, or at rest after the cause is eliminated. The presence of two of five components suggests the atypical angina, and one or no sign indicates the extracardiac pain. The atypical angina is also diagnosed in case of neck, jaw, tooth, back, hand pains, if other signs support this hypothesis.

As one can see, CHD can be diagnosed basing on the pain syndrome only in case of typical angina. However, it accounts for only 30% of established angina, the other 70% being its atypic counterpart. All cases with 10-90% "pre-test probability" require diagnostic tests, which help to reveal the anatomic substrate of the disease (usually the coronary atherosclerosis) and the presence of the myocardial ischemia [29-31]. The anatomy of the coronary arteries is assessed by electron beam computed tomography (EBCT) and multi-detector or multi-slice CT (HDCT) or invasive coronary angiography. The main functional tests for the ischemia are performed by ECG, myocardial perfusion scintigraphy with SPECT, stress echocardiography, and stress magnetic resonance imaging. Each of the techniques has its sensitivity and specificity, predictive value, and precision, which is different in different groups of patients.

For example, the ECG interpretation in exercise tests differs in men and women, the old and the young, patients with and without diabetes mellitus. The false-positive rate of ECG treadmill exercise testing is higher in women (38% to 67%) than men (7% to 44%), and the false-negative rate is low [29]. The relationships between the angina and the signs of ischemia on ECG are very complicated if ever exist. Heart and Soul Study (HSS) gave the evidence that 75% of patients with established changes in coronary arteries have angina attacks with no diagnosed ischemia, and 25% cases of diagnosed ischemia have no attacks of angina [32].

There is also a serious problem of silent ischemia. It was revealed that 28% male and 35% female patients had silent, myocardial infarction [33], 51% patients with diabetes mellitus, dyspnea and no chest pain had objective evidence of coronary artery disease (CAD) by SPECT criteria [34]. The positive predictive value for detecting CAD by coronary angiography in patients with silent ischemia ranged between 60% and 94% and was higher in men than women [35].

The above mentioned reasons substantiate the use of additional diagnostic methods, including artificial neural networks.

When one uses the techniques based on artificial intelligence, it should be clearly understood that the machine solves the task and asks the questions formulated by a researcher. We can suppose that the precision of the answers depends on many factors connected with the quality of the ANN, the characteristics and the preparation of the input data. In this work we will consider two problems, which, in our opinion, are of practical interest: the selection of initial set of signs describing the object, and the optimal type of ANN algorithm.

Network Algorithm

Usually, much attention is paid to data mining methods and algorithms. In fact, there is a range of approaches to solve the assigned tasks; the most popular are support vector machines and multilayer perceptrons with various algorithms for better training. Because these network topologies are well known, here we provide only short descriptions of some algorithms.

Support Vector Machine (SVM) solves a problem of binary classification by the construction of a hyperplane which divides objects into 2 classes. The method is based on a non-linear transformation of input features into the space of their linearly discriminabile separable images (K (x, x) of a higher dimensionality. Among the advantages of SVM networks is the possibility of calculation of the number of processing elements (neurons) in a hidden layer by an algorithm based on the number of support vectors without without additional settings. However, due to its computational complexity, the training of the algorithm takes a long time.

Multilayer Perceptron (MLP) neural network is a feedforward network consisting of input layer neurons, one or more hidden layers and one output layer. The training process is often based on a back propagation of errors algorithms, where all synaptic weights are adjusted to bring an output closer to the desired signal. Program is set in accordance with the rule of error correction based on the differences between a desired (target) response and an actual output.

MLP is the most widely-used neural network topology in medicine as well as other fields. Generalizing properties of this type of networks are well-studied and tested; their advantages include developed training methods, which are not limited to back propagation. There are also correlations which enable to assess a complexity of a network needed for a certain number of examples in a training set. A disadvantage of this topology is that a specific MLP network should be constructed for each problem using heuristic algorithms.

*Radial basis function***.** Network topology based on radial basis function (RBF) requires three layers with different functions. An input layer consists of input neurons connected to a network environment. Neurons of the second (hidden) layer perform non-linear transformation of the input into a hidden space of higher dimensionality to enhance its linear separable, as we noted above (see SVM description). Hidden layer neurons perform a set of functions which serve as an arbitrary basis for decomposition of input vectors. The number of neurons in output layer is equal to the number of classes to be defined. Since most RBF networks perform transformation using Gaussian function, a local approximation of non-linear mapping is created. In theory, any transformation by an RBF network can be performed by MLP network with the same level of accuracy. The advantage of RBF network is relative easiness of adjustment. On the other hand, a greater number of hidden layer neurons is needed for the mapping by Gaussian functions. This limits the use of RBF networks to problems with high-dimensional input space.

Modular networks consist of several neural network subsystems, which independently process various input signals. The output is then integrated into a single module, which determines an output signal and a set of examples to be used in training of specific subsystems. Disadvantages of this typology include its complexity and difficulties with adjustment and training. The advantage is the possibility to improve the efficacy of the whole system by specialized training of specific subsystems.

Fuzzy Logic Networks (FLN) give the ability to provide unknown rules of input mappings to output in various systems. Moreover, any non-linear function with several variables can be represented as a sum of fuzzy functions of one variable. This topology is commonly used in combination with modular networks (Intelligent Heart Disease Prediction System with CANFIS and Genetic Algorithm). In practice, such network combinations provide better results compared to hybrid networks based on Kohonen algorithm.

Hybrid networks based on genetic algorithms, self-organization algorithms and principal components method. Hybrid networks are a powerful tool that combines self-organization neural networks and networks with back propagation training (training with a teacher) in a single complex. Hybrid networks are able to divide a given set of data points into homogeneous clusters with a clustering algorithm, which allows using a simple topology in a part of network to solve the problem with a teacher, and thus facilitating the training process.

Genetic algorithms (GA) is an adaptive search method for functional optimization problems. They are based on genetic processes in biological organisms. GA randomly generates an initial population of structures and then works iteratively until it reaches a defined number of generations or some other stop criterion. In each generation, selection is performed in accordance to a fitness function, which is defined by an objective function to be minimized. The following processes are performed according to biological signs:

- Selection — a process of choosing a vector to form the next generation of fitness function values;
- Crossing — a process of selection the most adapted vector (the vector with the lowest values in objective function);
- Mutation — a process of replacing some value of an element of a vector with a valid randomly selected value.

Algorithms of self-organization. Self-organization is a process of spatial, temporal, or space-time ordering in an open system by coherent interactions between many elements of its components. As a result, a unit of next quality level appears. Methods of self-organization are based on Kohonen self-organizing maps that transform an input space of features into a discrete output while preserving the topological properties of the input. These methods are also used for data clustering in order to reduce dimensionality of input data, which justify their use in hybrid networks.

Principal component method. This principal component method reduces dimensionality of input data in order to find out a substantial part of the information. The algorithm is based on representation of input space as a sum of mutually orthogonal projection and discarding linear combinations of symptoms with low dispersion. The method can reduce the dimensionality of input data, which simplifies the topology of neural network and facilitates a training process.

When talking about accuracy of ANN models, it should be understood that it is influenced by multiple factors. Among them are all the factors connected with the process of ANN learning, such as representativity of the learning set, algorithm settings, etc.

Input and Reference Signs

Selection of Input Signs

Any standard technique of intellectual data analysis recommends selection and preparing the input data of the highest possible quality for further transformations and model construction. This crucial process takes about 70% of the time and effort. Usually, such data are relatively random, and may depend on a problem to solve, or a formulated hypothesis, or our concept of the features of an object (ex. patient, ECG, cardiac imaging), etc. If a disease is to be diagnosed with an ANN, the database must contain descriptions of the objects with a sufficient number of input signs closely connected with the disease; the database should also include the required output sign, which corresponds to the diagnosis. All these conditions are very important for an adequate setting and learning of the network. To achieve success, we should also understand the essence of each sign and its role in the solution of a medical problem.

Objective difficulties with the CHD detection occur due to multiplicity and variability of data to be taken into account. A modern practicing cardiologist should be able to possibly quickly assess (i.e. to analyze and correctly interpret) an anamnesis, signs and symptoms, the results of a number of laboratory and instrumental tests (ECG, doppler and echocardiography, coronary angiography, scintigraphy at rest and after exercise, etc.). The overall number of quantitative and qualitative signs may be as high as several hundred. On initial stage they are usually classified by source or some other principle. The most popular types of sign are demographic and clinical data, and CHD markers determined by laboratory and instrumental tests. Another classification principle involves grouping together substantially similar data, which reflect the same pathology and therefore provide similar clinical information. The examples are total cholesterol, LDL and triglyceride level for the dyslipidemia; echocardiography measurement of left ventricular wall thickness and ECG measurement of the amplitude of the waves of QRS complex for the myocardial hypertrophy, scintigraphic detection of perfusion defect of myocardium and ECG tracing of changes in QRS complex, ST segment and T wave for the myocardial ischaemia, myocardial infarction etc. Another approach is to separate the original data and the data derived from them. For instance, body mass index (BMI) is secondary information calculated from the body mass and height; atherogenic index is derived from the cholesterol fraction; left ventricular ejection fraction is calculated from the left ventricular end-diastolic and end-systolic dimensions etc. This explains why redundant signs may be included into a model and cause its subsequent sophistication. Moreover, each new sign increase the probability of inclusion of unreliable, distorted of incomplete values. Most neural networks are robust for such situations, however if they go beyond a certain limit, the accuracy of the result may be affected. To avoid this, the initial number of input signs should be controlled and probably reduced.

The set of input signs depends on the strategy we choose to diagnose or predict CHD. In case of image analysis, the set of signs is seemingly determined by the diagnostic technique. This is true, but in fact the process is significantly influenced by an assigned task. For example, ECG detection of myocardial ischemia usually includes quantitative assessment of ST segment and T wave detected in the 12-Lead ECG (interval between the ST-J point, ST-J amplitude, ST slope, ST amplitude, positive T amplitude, negative T amplitude)[36-41]. If the task is to classify arrhythmias basing on ECG values, other characteristics of ECG signal are taken into account, namely the amplitude and duration of QRS the P, PP и RR intervals between two successive P or R waves etc. Sometimes, multiple ECG signal data points merge into a few represent parameters most important for recognition and diagnostic purposes. A lot of studies have been done and good results achieved in respect of the detection of CAD by different diagnostic techniques, since ANNs have been applied for image analysis in heart disease diagnosis. Table 1 shows some results of ANN-enhanced diagnosis of myocardial ischemia using various techniques. As one can see, ECG and myocardial perfusion scintigraphy at rest and after exercise are the topic of most of these works. This may come from the fact that the methods give a relatively high discordance in data interpretation and quite a lot of false conclusions; ANN application helps to improve the situation in spite of the differences in accuracy among models and even within one model. However, it should be noticed that ANN models can provide equal sensitivity in different diagnostic tests (ex. 90% in ECG and myocardial scintigraphy as well).

Interestingly, in case of the coronary artery disease there is a discrepancy in diagnostic accuracy between ANN and experts, which depends on the character and localization of the pathologic process. For instance, ANN diagnosis is more accurate in case of single-vessel compared to multivascular disease [60]. Similarly, myocardial perfusion scintigrams of the right coronary artery infarction are more correctly interpreted by ANNs, while the reversible ischemia in the left coronary artery is better detected by experts [42].

Diagnosis and predictions based on sign complexes are rarely performed using one type of data only (clinical or laboratory-instrumental). Usually, the groups are combined in different ways. Input signs of the heart disease can vary significantly depending on the problem to solve. This leads to various models of similar or different informativeness. For example, M. Ture et al. generated a prognostic ANN model for arterial hypertension with 85.54% prediction rate using 10 signs: hypertension, smoking habits, lipoprotein (a), triglyceride, uric acid, total cholesterol, body mass index etc [51]. S. Huang et al. approached the same task with another set of signs, which considered occupation, family history, educational level, alcohol intake, vegetable and fruit intake, salty diet, animal insides intake, physical exercise, BMI, blood pressure difference and other. The researchers employed an MLP with the standard backpropagation algorithm and achieved 90% accuracy [52].

Table 1. The sensitivity and specificity of ANNs in the diagnosis of myocardial ischemia by different techniques

Diagnostic technique	Publication	Analyzed object	Sensitivity(%)	Specificity(%)
Myocardial perfusion scintigraphy	[42]	perfusion and functional image data	77.2	77.2
ECG	[43]	QRS complex, ST segment and the T wave	79	75
Myocardial perfusion scintigraphy	[44]	perfusion and functional image data	correct classifications in 71%	
Myocardial perfusion scintigraphy	[45]	perfusion and functional image data	72 - 74	73 - 77
ECG	[46]	QRS complex, ST segment and the T wave	81	84
ECG	[47]	QRS complex, ST segment and the T wave	90	90
Myocardial perfusion scintigraphy	[48]	perfusion and functional image data	90	85
Echo	[50]	waveforms strain rates and strains or of time intervals during selected phases of the cardiac cycle	86	87

When ANNs are applied in coronary artery disease, the variety of input signs can be even higher. The number of signs may differ significantly, usually up to several dozens. In the work by R.F. Harrison et al. more than 40 potential input variables were used to predict acute coronary syndromes [53]. Some models emphasize laboratory data [54, 55], other [56; 57] pay more attention to exercise tests. Many researchers combine demographic characteristics, anamnesis and the results of various diagnostic techniques [56, 58]. A.M. Bulgiba and M. H. Fisher used 94 signs to diagnose acute myocardial infarction, including demographic, character of chest pain, cardiac risk factors, and general examination of associated heart/lung symptoms [59].

Selection of Reference Signs

It should be noticed that the result of a supervised learning of CHD diagnosis is significantly influenced by the reference sample and the criteria of normality. This was shown by J. Toft et al. [67] who compared the accuracy of networks that were taught CHD classification using myocardial perfusion scintigrams obtained with different reference standards: one group consisted of patients with normal coronary angiography, and another enrolled healthy volunteers with <5% likelihood of CAD. The use of first group resulted in 93% area under ROC curves, 80% sensitivity and 87% specificity, compared to 72% area under ROC curves and 50% sensitivity in case of the group of healthy persons. Actually, the choice of a reference sample is directly connected with the type of diagnostic test. Coronary arteriography seems to be ideal for modeling purposes if its invasiveness is not taken into account, and if it is performed on patients for certain indications. Because of these restrictions, the number of patients with healthy coronary arteries is often insufficient, which in turn decreases the quality of model. However, the intention to simplify the database building should not be the reason for choosing some intermediate diagnostic criteria. This may cause missing of the data necessary for the construction of an adequate model. Nevertheless, coronary arteriography now remains the golden standard for CHD diagnosis. However, its application as a reference technique encounters some difficulties because of different standards used for the interpretation of the results. Usually, yes/no classification of coronary heart disease considers ≥ 50% narrowing of main coronary arteries; sometimes CAD is defined as ≥ 60% or 75% stenosis [60; 67]. Different network accuracy is obtained when stenosis localization is added as an additional sign. For example, ANN models developed by Allison JS et al. [60] can predict coronary artery disease basing on stress single-photon emission computed tomographic images with 96% accuracy in single-vessel involvement, but only 65% in case of multivessel involvement. The sensitivity of models for coronary artery disease is within the range of 69% (left circumflex artery) to 92% (left anterior descending artery) with 93% and 78% specificity, respectively. One has to agree with H. Haraldsson that the stenosis in a coronary artery does not always correlate with a reduction in myocardial perfusion or ST depression [63].

Table 2. The results of ANN application in coronary heart disease

Publication	Object	Sample size	Reference standards	ANN topology	Input data	Precision
Allison [60]	coronary artery disease	109 patients	coronary angiography	multilayer perceptron	25 data points from myocardial perfusion images	96% accuracy for single-vessel CAD, and 65% for multivessel CAD
Atkov [61]	coronary artery disease	487 patients	coronary angiography	multilayer perceptron	14 candidate gene polymorphisms, 18 non-genetic CHD risk factors: age, gender, total cholesterol, high-density lipoprotein cholesterol, low-density lipoprotein cholesterol, very-low-density lipoprotein cholesterol, triglycerides, cholesterol ratio, fasting plasma glucose, arterial hypertension, diabetes mellitus, current tobacco smoking status, BMI, a family history of CHD, profession, SCORE index, left ventricular ejection fraction, coronary angiography data	94% accuracy
Chen [62]	coronary artery disease	2949 patients	coronary angiography	Bayesian networks, naive Bayes, support vector machine, k-nearest neighbor, neural networks decision trees	10 non-genetic risk factors, 30 candidate gene polymorphisms	77,2-89,7% sensitivity 86,5-88,9% specificity 81,9-89,2% accuracy

Publication	Object	Sample size	Reference standards	ANN topology	Input data	Precision
Çolak [9]	coronary artery disease	124 patients	coronary angiography	multilayer perceptron	17 non-genetic input variables: sex, age, hypertension, diabetes, family history, smoking, stress, physicalactivity, BMI, hemoglobin, white blood cells, uric acid, triglycerid, HDL, LDL, direct and total bilirubin	71-96 % sensitivity 76-94%, specificity 78-92% accuracy
George [55]	coronary atheroscle rosis	81 patients	coronary angiography	multilayer perceptron	20 input variables: hypertension, oxidized and native LDL, phosphatidylserine, phosphatidylcholine, phosphatidylethanolamin, serum albumin, homocystein, C-reative protein, β2-glycoprotein I, anticardiolipin, cytomegalovirus, diabetes mellitus, herpes simplex virus 1 and 2, Chlamydia pneumonia, Helicobacter pylori	70% sensitivity 80% specificity 78% accuracy
Haraldsson [63]	coronary artery disease myocardia l ischemia	229 patients	coronary angiography	multilayer perceptrons network by Bayesian learning	30 values describing the myocardial perfusion images, 11 exercise test data (heart rate, workload, ST60 amplitudes)	0.78 ROC area for CAD 0.88 for ischemia

Table 2. (Continued)

Publication	Object	Sample size	Reference standards	ANN topology	Input data	Precision
Karabulut [64]	coronary artery disease	303 patients	coronary angiography	Multilayer neural networks with Levenberg-Marquardt algorithm, Rotation Forest ensemble method	14 input variables: age,sex, chest pain type, resting systolic blood pressure, serum cholesterol, fasting glucose, resting ECG, maximum heart rate, exercise induced angina, ST depression, slope of the peak exercise ST segment, number of major vessels colored by fluoroscopy, exercise thallium scintigraphic defects	65.9-85% sensitivity 71.5-86.7% specificity 71.62-85.14% accuracy
Lapuerta [54]	Risk of coronary artery disease	188 patients	occurance of coronary events: death, myocardial infartion, angioplasty	multilayer perceptron	7 input variables: cholesterol, HDL, LDL, Triglyceride, ApoB, ApoCHP, ApoCR	66% success rate
Lindahl [45]	coronary artery disease	135 patients	coronary angiography	multilayer perceptron	30 values describing the myocardial perfusion images, 2 gender variables	92 -98% sensitivity 62-81% specificity
Mobley [8]	coronary artery stenosis	763 patients	coronary angiography	multilayer perceptron ROC analysis and logistic regression	14 input variables: age, sex, race, smoking current, diabetes, hypertension, BMI, creatinine, triglycerides, cholesterol, HDL, cholesterol: ratio, fibrinogen, lipoprotein(a)	0.89 ROC area

Publication	Object	Sample size	Reference standards	ANN topology	Input data	Precision
Scott [56]	coronary artery disease	102 patients	coronary angiography	neural network	20 input: clinical, treadmill exercise tests and myocardial perfusion imaging data	88% sensitivity 65% specificity
Srinivas [57]	coronary artery disease	-	coronary angiography	multilayer perceptron decision tree neuro-fuzzy network Bayesian Network support vector machine	sex, chest pain type, fasting blood sugar, resting ECG results, exercise induced angina, slope of the peak exercise ST segment, number of major vessels colored by floursopy, Thalium defect, blood pressure, cholesterol, maximum heart rate achieved, ST depression induced by exercise relative to rest, age	87-90.17% sensitivity 82- 89.7% accuracy
Stefko [65]	coronary artery disease	580 data records	coronary arteriography	multilayer perceptron	traditional ECG exercise test data	96% accuracy
Tham [66]	coronary artery disease	704 patients	coronary angiography	multilayer perceptron, single-layer Hierarchical Mixture of Experts (HME), multi-layer HME	19 genetic and 10 non-genetic (race, sex, smoking habits and family history, total cholesterol, HDL, LDL, triglycerides, BMI, age)	74.76 - 87.32% accuracy
Toft [67]	coronary artery disease	87 patients and 128 healthy volunteers	coronary angiography; likelihood for CAD <5%	multilayer perceptron	30 values describing the myocardial perfusion images	80% sensitivity 87% specificity

This discrepancy appears because the stenosis highlights an atherosclerotic plaque as a morphological substrate of the disease, while perfusion defections or changes in ST segment are a sign of myocardial ischemia. The same problem may emerge, if ANN models are enhanced with other methods of coronary atherosclerosis detection, for example single photon emission computed tomography (SPECT).

Based on the above, the difficulties with the comparison of model accuracy in different reference methods and standards become obvious. With this consideration, the requirements for the reference methods should be quite rigid if we want to obtain networks with adequate diagnostic and prognostic capabilities. It is reasonable to propose using two reference methods in place of one, for example coronary angiography and myocardial scintigraphy, or coronary angiography and ECG.

Table 2 summarizes the results of several studies on ANN application in coronary heart disease. The studies are methodologically different and therefore cannot be directly compared; however, we can conclude, that MLP topology provides the best results in most cases.

Examples of the Application of ANNs in the Diagnosis of Coronary Artery Disease

This section presents the results of study aimed to develop an ANN model for the coronary heart disease. The objectives were to analyze the methodological aspects of modeling with unequal sample of patients and healthy persons, the selection of significant signs on the basis of hierarchical clustering, the influence of a big group of signs on the disease detection, and the diagnostic accuracy in different sets of signs and network topologies.

Materials and Methods

The study included 279 sequentially selected patients (245 males, 34 females, mean age 53.9) hospitalized for a coronary angiography to diagnose CHD. All patients underwent standard clinical examinations (laboratory tests, ECG, Holter monitoring, stress tests, echocardiography etc.), genetic analysis, and coronary angiography. The diagnosis of CHD was made by a physician after both ECG and coronarography. Coronary atherosclerosis was angiographically diagnosed in 235 (84.2%) patients, and 44 (15.8%) patients had an intact coronary artery wall. The criterion for normal coronary arteries was the absence of atherosclerotic plaque in major epicardial arteries. The information obtained from testing and genotyping allowed us to create a database of patients that was subsequently used to diagnose coronary heart disease by ANN.

Data Collection and Processing

The set of variables consisted of 60 CHD signs: demographic, anamnestic, laboratory, ECG, EchoCG, and genetic (see Table 3).

Table 3. Classification of input variables

Groups	Variables
Demographic	age, gender, profession
Cardiac risk factors	family history of CHD, diabetes mellitus, current tobacco smoking status, SCORE index
Symptoms	chest pain
Physical examination	weight, height, BMI > 30, systolic and diastolic blood pressure, heart rate
Laboratory tests	total cholesterol, HDL cholesterol, LDL cholesterol, VLDL cholesterol, triglycerides, cholesterol ratio, fasting plasma glucose, hemoglobin, red blood cells, white blood cells, ESR, urea, creatinine, ALT, AST, LDH, CK-MB, serum potassium, sodium
ECG	pathologic Q wave, T wave, ST depression
EchoCG	interventricular septal and left ventricular (LV) posterior wall thickness, LV end-diastolic and systolic size and volumes, LV ejection fraction
Genetic	17 SNPs localized in genes: lipoprotein lipase (LIPC-250G/A) and LIPC-514C/T, nitric oxide synthase (NOS E298D), methylenetetrahydrofolate reductase (MTHFR A223V), angiotensin-converting enzyme (ACE Alu Ins/Del I>D), angiotensinogen (AGT M235T and AGT T174M variants), angiotensin II type 1 receptor (AGTR A1166C), plasminogen activator inhibitor-1 (PAI-1 5G/4G), and C-reactive protein (CRP-1, CRP-2, CRP-3, CRP-4, and CRP-5, circadian locomotor output cycles kaput (CLOCK - 3111 T/C, period homolog 1 (PER1 2434 T/C, period homolog 2 (PER2 111C/G)

The sets of variable parameters were selected to adjust ANN models by the pairwise correlation between the parameters and CHD diagnosis. The task to solve was of two-class classification: “1” (CHD) or “0” (healthy). 209 examples were used for training, and 70 for testing. *NeuroSolutions 5.0* environment was used to check the possibilities of optimization.

The process of CHD diagnosis has some specific features connected with the obtaining of clinical data: the number of patients is sufficiently higher than the number of healthy persons, and the number of cases that can be included into a study is usually limited. Only nine of 209 included patients were healthy.

This made necessary to decrease the number of signs and to select most informative ones, and on the other hand, special techniques should be applied to minimize the disproportions arising because of the different contribution of patients and healthy volunteers to the learning sample.

The limited number of cases imposes the use of ANN models with a lower number of settings according to the formula: $\frac{W}{\varepsilon} \sim N$, where W is the number of settings, N is the size of a learning sample, and ε is the learning error.

To cope with the different contribution of sample groups, the following techniques were applied:

- WeightGradient. An approach, in which the contribution of a single case is multiplied by a coefficient, which is determined for each sign and stored in a separate file.
- Special enrichment of the learning sample by „healthy” cases created by a computer generation of input values laying in a range acceptable for a healthy person.

The number of signs was reduced by as following. Pair correlation was calculated for initial (input) signs; the pairs of signs with maximal correlation were of the most interest.

On the next step, the parameters were clustered to reduce the number of signs. To do this, after the pair correlation ($\rho(\vec{x},\vec{y})$) was calculated, a measure of relationship was introduced for $\vec{x},\vec{y}$ from the formula

$$d(\vec{x},\vec{y}) = \begin{cases} 1-|\rho(\vec{x},\vec{y})|, |\rho(\vec{x},\vec{y})| > 0.4 \\ 1, |\rho(\vec{x},\vec{y})| \le 0.4 \end{cases}.$$

A similarity matrix was then constructed, and a clustering was performed on its basis. As a result, the groups of most closely correlated signs were selected (see the tree diagram on the Figure 1). This hierarchical clustering allowed separating the data set into correlated and non-correlated category.

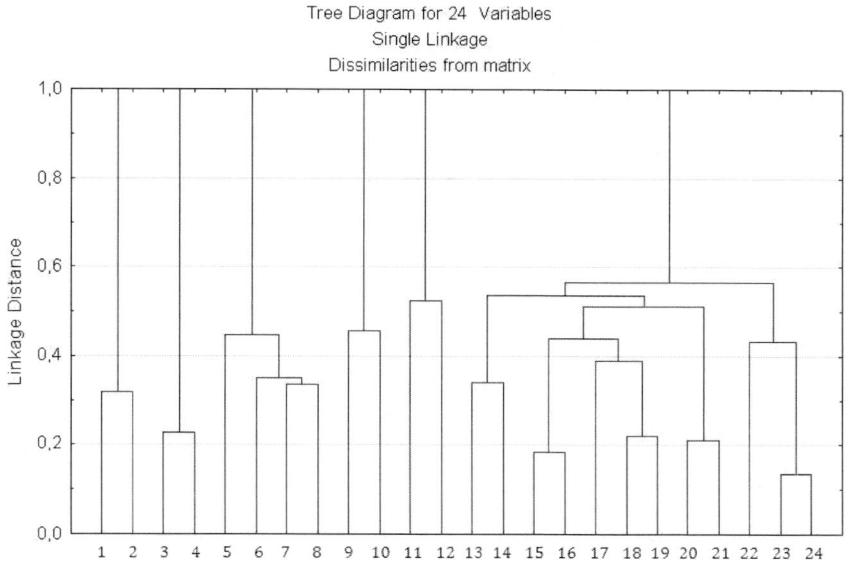

Figure 1. Clustering of the signs with maximum correlation. 1. LV end-diastolic size 2. LV end-systolic size 3. serum potassium, 4. serum sodium 5. LDH 6. CK-MB 7. CK 8. AST 9. fasting plasma glucose 10. diabetes mellitus 11. left ventricular ejection fraction 12. pathologic Q wave 13. interventricular septal thickness 14. LV posterior wall thickness 15. triglycerides 16. very-low-density lipoprotein (VLDL) cholesterol 17. cholesterol ratio 18. low-density lipoprotein (LDL) cholesterol 19. total cholesterol 20. systolic blood pressure 21. diastolic blood pressure 22. weight 23. height 24. BMI.

Next, the correlation between the input signs in the learning subset and the intended output was calculated.

The best correlated inputs and outputs were: BMI, pathologic Q wave, systolic and diastolic blood pressure, red blood cells, total cholesterol, very-low-density lipoprotein (VLDL) cholesterol, interventricular septal and LV posterior wall thickness, left ventricular ejection fraction.

The following learning sets were prepared (see Table 4):

Table 4. Learning sets and sample groups

	Learning set 0	Learning set 1	Learning set 2	Learning set 3
Number of signs	all signs	all signs	all non-correlated signs and one sign from the clusters presented on the Figure 1 (LV end-systolic size, serum sodium, LDH, CK, fasting plasma glucose, diabetes mellitus, left ventricular ejection fraction, systolic blood pressure, BMI).	input signs, which were the best correlated with the intended output (correlation coefficient > 0.4).
Sample group	9 healthy persons, and 200 patients correction by weight gradient learning algorithm	209 healthy persons and 209 patients (200 additional healthy persons were created artificially by the generation of values selected for diagnosis of paramcters in acceptable range for healthy patients)	209 healthy persons and 209 patients (200 additional healthy persons were created artificially by the generation of values selected for diagnosis of parameters in acceptable range for healthy patients).	209 healthy persons and 209 patients (200 additional healthy persons were created artificially by the generation of values selected for diagnosis of parameters in acceptable range for healthy patients).

ANN Algorithms

Several models were created. A basic neural network model was created using a multilayer perceptron (MLP) with two hidden layers enhanced with a genetic algorithm configured to calculate the number of neurons (computing units) in the hidden layers. It was used in combination with the following neural networks:

- Support vector machine (SVM);
- Probabilistic neural network (PNN);
- Hybrid neural network: radial basis function (RBF) networks combined with MLP with one hidden layer. The genetic algorithm was configured to choose the parameters of RBF network and the number of neurons in the hidden layer;
- MLP network with two hidden layers and input processing by principle component analysis (PCA). The number of neurons in hidden layers and the number of principle components was determined by a genetic algorithm.

Testing

To diagnose CHD, all four learning sets (0-3) were processed by every network. Testing sample consisted 35 patients and 35 healthy persons.

Results

Training of the selected networks with the set 0 provided no meaningful result. This means that the diagnostic accuracy may be significantly improved by computer generation of input values appropriate for healthy persons, if there are considerable differences in the number of patients and healthy persons in the learning set. The result was also unsatisfactory when the 3rd learning set was used. It seems that the use of inputs with maximum correlation with the intended result does not give a full and complete disease description. Some hidden relations between groups of signs may influence the result of the classification, and they probably were not taken into account during the preparation of the learning set.

Table 5 presents the results of learning and testing for the sets 1 and 2. Figure 2 shows ROC curves for each neural network.

As one can see, the accuracy of the models varies among different ANN types and sample sets. It amounted to 57–77% in the first, and 51–80% in the second set. The best result (80%) was achieved in case of the MLP network with two hidden layers and genetic algorithm with hierarchic clustering of input parameters. The PNN showed the least satisfactory result in both learning sets (57 in the first and 51 in the second).

These values were compared with our previous study [61], which gave a better result: 93% accuracy of CHD diagnosis in case of a set of 8 SNPs, SCORE index and coronary angiography data. This difference in accuracy may result from the fact that the previous work was based on other groups and sets of signs, and the initial learning set had a sufficient healthy/patient proportion (2:1). There was no need in artificial increase of the number of healthy persons. Additionally, an excessively large number of signs appear to play rather negative role. One of the conclusions is that the optimal number of input parameters should not exceed 8–10 most crucial factors. The use of all parameters seems to complicate a model, while lower numbers does not provide sufficient information to solve a classification task.

Table 5. Accuracy of ANN models for CHD diagnosis

Neural network	Learning set 1		Learning set 2	
	Learning	Testing	Learning	Testing
MLP with two hidden layers and genetic algorithm*	98%	57%	94%	80%
SVM	100%	68,5%	100%	73%
PNN	100%	57%	100%	51%
Hybrid RBF network combined with an MLP network with one hidden layer and a genetic algorithm	69,8%	70%	73,5%	74%
MLP network with two hidden layers and data processing by principle components analysis (PCA).	100%	77%	99,7%	77%

* with hierarchical clustering of input parameters.

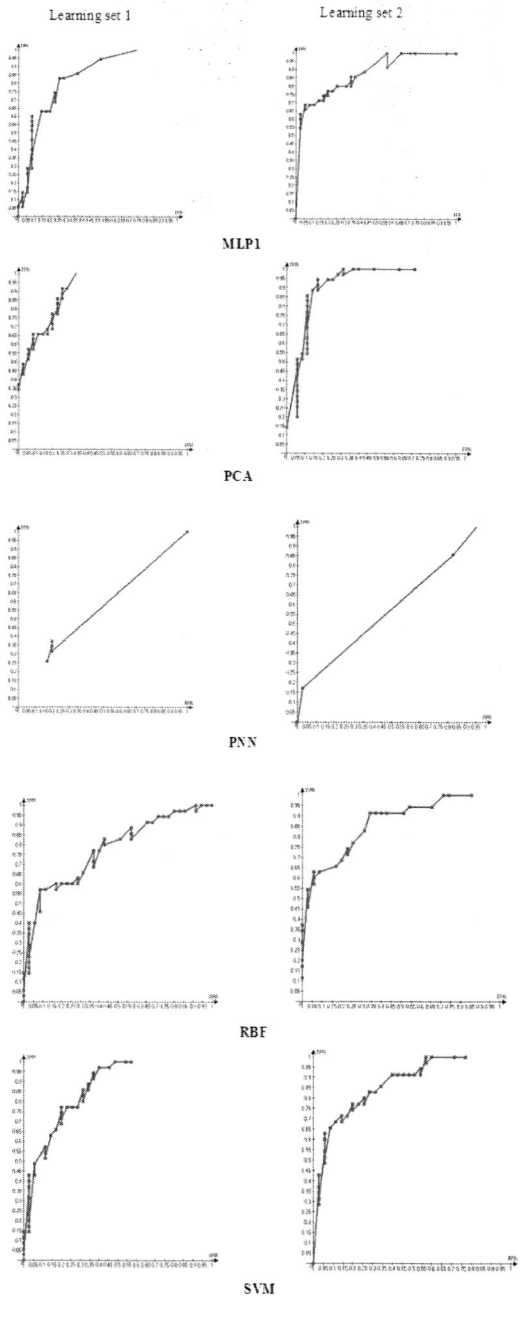

Figure 2. ROC Curves of neural networks for CHD (learning sets no. 1 and 2).

The achieved accuracy values are generally similar to those in studies, where a larger number of signs were used [9, 55, 66]. A higher accuracy was only reached in [62]; the model included 40 signs, however the proportion of non-genetic factors and gene polymorphisms was 1:3 compared to 3:1 in our work. In this regard, it should be noticed that the specificity of signs, which was determined in our work, clarified the significance of several signs in CHD diagnosis and confirmed previously known role of blood lipid level, professional factors and some genes.

CONCLUSION

In this work we only dwell on the issue of the use ANN in CHD diagnosis. In fact, the field of possible ANN applications in cardiology is much broader. It may be used to diagnose and predict arrhythmia [68-70], heart valve disease [71, 72], cardiac surgery [73-75], cardiovascular medications [76] etc. Obviously, artificial neural networks cannot fully substitute a physician, but they can help to accelerate and simplify the choice of the solution. Exactly as in case of the human brain, their efficiency depends on the precision of assigned task, on how fully the object of study is described, and on the quality of ANN learning. Our results suggest that there is no need in excessive amount of input data, because redundant information decreases the precision of ANN. About ten most important signs seem to be enough, but they should possibly include signs of different type; an optimal set should compose clinical laboratory and genetic data. This is confirmed by the selection of significant signs by hierarchy clustering (Figure 3). For example, in case of CHD such set may include systolic blood pressure, BMI, cholesterol rate, fasting plasma glucose, left ventricular ejection fraction, profession and some genes connected with the development of the disease. There are more and more evidence that various genes play a role in CHD pathogenesis. This work includes several new polymorphisms of circadian genes (CLOCK - 3111 T/C, PER1 2434 T/C, PER2 111C/G), which have not been tested before. We are going to assess some other candidate genes from the point of their contribution to the ANN accuracy. In our opinion, the genetic component is indispensable, though it is premature to say what specific polymorphisms should be included.

The prominent fact is that no meaningful result was achieved, if the learning set was too small and the number of healthy persons and patients was unequal. We have intentionally analyzed such situation, because it is probable in clinical modeling.

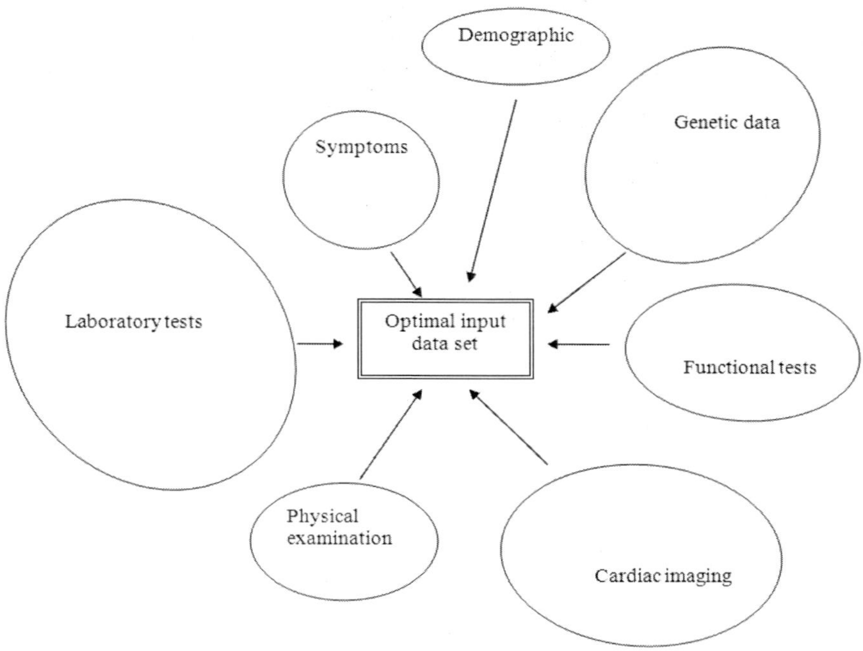

Figure 3. The selection of signs important to achieve a model with high diagnostic and prognostic characteristics.

For example, if a diagnostic test is performed on a patient according to strict medical conditions, it takes a long time to collect enough data because of the lack of healthy cases. The solution may be found in artificial replenishment of learning sets. To do this, we generated the values selected for diagnosis of parameters in acceptable range for healthy people, and thereby equated the number of patients and the healthy. This lead to significant increase of prognostic accuracy; yet, the result was somewhat inferior compared to the set with the real data.

It is of interest, that in this and previous work the accuracy was influenced by the type of ANN, although to a lesser extent. The best diagnostic precision was achieved in case of MLP with two hidden layers and genetic algorithm. A similar result was shown in case of SVM, MLP network with two hidden layers and data processing by principle components analysis and hybrid RBF network combined with an MLP network with one hidden layer and a genetic algorithm. On the other hand, the accuracy of probabilistic neural network (PNN) was significantly lower. It should be noted that no significant differences in accuracy was found between the learning and testing sample in hybrid RBF network combined with an MLP network with one hidden layer and a genetic algorithm; all other ANN types gave excellent results in the learning set, but not in testing set. All this shows that the result can be optimized by the appropriate selection of ANN algorithm, but to achieve significant improvement, fundamentally new types of networks must be developed.

REFERENCES

[1] Suzuki, K. Artificial Neural Networks - *Methodological Advances and Biomedical Applications.* 2011.

[2] Anagnostou, T; Remzi, M; Lykourinas, M; Djavan, B. Artificial neural networks for decision-making in urologic oncology. *Eur. Urol.* 2003 Jun 43(6), 596-603.

[3] Lisboa, PJ; Taktak, AF. The use of artificial neural networks in decision support in cancer: a systematic review. *Neural Networks.* 2006 May 19(4), 408-415.

[4] Dietzel, M; Baltzer, PA; Dietzel, A; Vag, T; Gröschel, T; Gajda, M; Camara, O; Kaiser, WA. Application of artificial neural networks for the prediction of lymph node metastases to the ipsilateral axilla — initial experience in 194 patients using magnetic resonance mammography. *Acta Radiol.*, 2010 Oct 51(8), 851-858.

[5] Chi, CL; Street, WN; Wolberg, WH. Application of artificial neural network-based survival analysis on two breast cancer datasets. *AMIA Annu. Symp. Proc.*, Oct 2007 11, 130-134.

[6] Abbod, MF; Catto, JW; Linkens, DA; Hamdy, FC. Application of artificial intelligence to the management of urological cancer. *Journal of Urology*, 2007 Oct 178(4), 1150-1156.

[7] Shen, Z; Clarke, M; Jones, RW; Alberti, T. A New Neural Network Structure for Detection of Coronary Heart Disease. *Neural Computing and Applications.* 1995 Jan 3. 171-177.

[8] Mobley, BA; Schechter, E; Moore, WE; McKee, PA; Eichner, JE. Predictions of coronary artery stenosis by artificial neural network. *Artificial Intelligence in Medicine.* 2000 18. 187-203.

[9] Çolak, MC.; Çolak, C.; Kocatürk, H.; Sagiroglu, S; Barutçu I. Predicting coronary artery disease using different artificial neural network models. *Anadolu Kardiyol Derg.* 2008 8. 249-254.

[10] Nahar, J; Imam, T; Tickle, KS; Chen YPP. Computational intelligence for heart disease diagnosis: A medical knowledge driven approach. *Expert Systems with Applications*. 2013 40. 96–104.

[11] Pace, F; Savarino, V. The use of artificial neural network in gastroenterology: the experience of the first 10 years. *European Journal of Gastroenterology and Hepatology*. Dec 2007 19(12). 1043-1045.

[12] Strategic priorities of the WHO Cardiovascular Disease Programme. Available from: *http://www.who.int/cardiovascular_diseases/priorities/en/index.html*

[13] US Department of Health and Human Services. HDS-2: Reduce coronary heart disease deaths. Healthy People 2020. Washington, DC: *US Department of Health and Human Services*. 2012.

[14] Roger; VL; Go, AS; Lloyd-Jones, DM; Benjamin, EJ; Berry, JD; Borden, WB; Bravata, DM; Dai, S; Ford, ES; Fox, CS; Fullerton, HJ; Gillespie, C; Hailpern, SM; Heit, JA; Howard, VJ; Kissela, BM; Kittner, SJ; Lackland, DT; Lichtman, JH; Lisabeth, LD; Makuc, DM; Marcus, GM; Marelli, A; Matchar, DB; Moy, CS; Mozaffarian, D; Mussolino, ME; Nichol, G; Paynter, NP; Soliman, EZ; Sorlie, PD; Sotoodehnia, N; Turan, TN; Virani, SS; Wong, ND; Woo, D; Turner, MB. Heart disease and stroke statistics — 2012 update: a report from the American Heart Association. *Circulation*. 3 Jan 2012. 125(1). e2-e220.

[15] Go, AS; Mozaffarian, D; Roger, VL; Benjamin, EJ; Berry, JD; Borden, WB; Bravata, DM; Dai, S; Ford, ES; Fox, CS; Franco, S; Fullerton, HJ; Gillespie, C; Hailpern, SM; Heit, JA; Howard, VJ; Huffman, MD; Kissela, BM; Kittner, SJ; Lackland, DT; Lichtman, JH; Lisabeth, LD; Magid, D; Marcus, GM; Marelli, A; Matchar, DB; McGuire, DK; Mohler, ER; Moy, CS; Mussolino, ME; Nichol, G; Paynter, NP; Schreiner, PJ; Sorlie, PD; Stein, J; Turan, TN; Virani, SS; Wong, ND; Woo, D; Turner, MB. Heart disease and stroke statistics — 2013 update: a report from the American Heart Association. Circulation. 2013. 12 Dec 2012. Available from: *http://circ.ahajournals.org/content/early/2012/12/12/ CIR.0b013e 31828124ad*

[16] European Cardiovascular Disease Statistics. 2012 edition. *European Heart Network and European Society of Cardiology*. Sept 2012.

[17] Berner, ES; Graber, ML. Overconfidence as a cause of diagnostic error in medicine. *Am. J. Med*. 2008 May 121(5 Suppl). S2-23.

[18] Tanindi, A; Cemri, M. Troponin elevation in conditions other than acute coronary syndromes. *Vasc. Health Risk Manag*. 2011 7. 597-603.

[19] Bösner, S; Becker, A; Abu Hani, M. Accuracy of symptoms and signs for coronary heart disease assessed in primary care. *Br. J. Gen. Pract.* 2010 Jun 60(575). e246-257.

[20] Bösner, S; Haasenritter, J; Abu Hani, M; Keller, H; Sönnichsen, AC; Karatolios, K; Schaefer, JR; Baum, E; Donner-Banzhoff, N. Accuracy of general practitioners' assessment of chest pain patients for coronary heart disease in primary care: cross-sectional study with follow-up. *Croat Med. J.* 2010 Jun 51(3). 243-249.

[21] Gencer, B; Vaucher, P; Herzig, L; Verdon, F; Ruffieux, C; Bösner, S; Burnand, B; Bischoff, T; Donner-Banzhoff, N; Favrat, B. Ruling out coronary heart disease in primary care patients with chest pain: a clinical prediction score. *BMC Med.* 2010 Jan 21 8:9.

[22] Bruins Slot, MH; Rutten, FH; van der Heijden, GJ; Geersing, GJ; Glatz, JF; Hoes, AW. Diagnosing acute coronary syndrome in primary care: comparison of the physicians' risk estimation and a clinical decision rule. *Fam. Pract.* 2011 Jun 28(3). 323-328.

[23] Verdon F, Herzig L, Burnand B, Bischoff T, Pécoud A, Junod M, Mühlemann N, Favrat B; GMIRG. Chest pain in daily practice: occurrence, causes and management *Swiss Med. Wkly*. 2008 Jun 138(23-24). 340-7.

[24] Berner, ES. Clinical Decision Support Systems: *Theory and Practice*. 2nd ed. 2007.

[25] Bellazzi, R; Zupan, B. Predictive data mining in clinical medicine: Current issues and guidelines. *Int. J. Med. Inform.* 2008 Feb 77(2). 81-97.

[26] Jiang, J; Trundle, P; Ren, J. Medical image analysis with artificial neural networks. *Comput. Med. Imaging Graph.* 2010 Dec 34(8). 617-631.

[27] Baxt, WG. Use of an artificial neural network for the diagnosis of myocardial infarction. *Ann. Intern. Med.* 1991 Dec 115(11). 843-848.

[28] Chest pain of recent onset: Assessment and diagnosis of recent onset chest pain or discomfort of suspected cardiac origin. NICE Clinical Guidelines, No. 95. London: Royal College of Physicians (UK). March 2010.

[29] Fraker, TD Jr; Fihn, SD; Gibbons, RJ; Abrams, J; Chatterjee, K; Daley, J; Deedwania, PC; Douglas, JS; Ferguson, TB Jr; Fihn, SD; Fraker, TD Jr; Gardin, JM; O'Rourke, RA; Williams, SV; Smith, SC Jr; Jacobs, AK; Adams, CD; Anderson, JL; Buller, CE; Creager, MA; Ettinger, SM; Halperin, JL; Hunt, SA; Krumholz, HM; Kushner, FG; Lytle, BW; Nishimura, R; Page, RL; Riegel, B; Tarkington, LG; Yancy, CW. 2007

chronic angina focused update of the ACC/AHA 2002 Guidelines for the management of patients with chronic stable angina: a report of the American College of Cardiology/American Heart Association Task Force on Practice Guidelines Writing Group to develop the focused update of the 2002 Guidelines for the management of patients with chronic stable angina. *Circulation*. 2007 Dec 4 116(23). 2762-2772.

[30] Fox, K; Garcia, MA; Ardissino, D; Buszman, P; Camici, PG; Crea, F; Daly, C; De Backer, G; Hjemdahl, P; Lopez-Sendon, J; Marco, J; Morais, J; Pepper, J; Sechtem, U; Simoons, M; Thygesen, K; Priori, SG; Blanc, JJ; Budaj, A; Camm, J; Dean, V; Deckers, J; Dickstein, K; Lekakis, J; McGregor, K; Metra, M; Morais, J; Osterspey, A; Tamargo, J; Zamorano, JL. Guidelines on the management of stable angina pectoris: executive summary. The Task Force on the Management of Stable Angina Pectoris of the European Society of Cardiology. *Eur. Heart J.* 2006 27(11). 1341-1381.

[31] Henderson, RA; O'Flynn, N. Management of stable angina: summary of NICE guidance. Guideline Development Group. *Heart.* 2012 Mar 98(6). 500-507.

[32] Gehi, AK; Rumsfeld, JS; Liu, H; Schiller, NB; Whooley, MA. Relation of Self-Reported Angina Pectoris to Inducible Myocardial Ischemia in Patients With Known Coronary Artery Disease: The Heart and Soul Study. *Am. J. Cardiol.* 2003 Sep 92(6). 705-707.

[33] Kannel WB. Silent myocardial ischemia and infarction: Insights from the Framingham study. *Cardiology Clinics*. 1986 4. 583-591.

[34] Zellweger, MJ; Hachamovitch, R; Kang, X; Hayes, SW; Friedman, JD; Germano, G; Pfisterer, ME; Berman, DS. Prognostic relevance of symptoms versus objective evidence of coronary artery disease in diabetic patients. *Eur. Heart J.* 2004 25. 543-550.

[35] Greenland, P; Alpert, JS; Beller, GA; Benjamin, EJ; Budoff, MJ; Fayad, ZA; Foster, E; Hlatky, MA; Hodgson, JM; Kushner, FG; Lauer, MS; Shaw, LJ; Smith, SC Jr; Taylor, AJ; Weintraub, WS; Wenger, NK; Jacobs, AK; Smith, SC Jr; Anderson, JL; Albert, N; Buller, CE; Creager, MA; Ettinger, SM; Guyton, RA; Halperin, JL; Hochman, JS; Kushner, FG; Nishimura, R; Ohman, EM; Page, RL; Stevenson, WG; Tarkington, LG; Yancy, CW. 2010 ACCF/AHA Guideline for Assessment of Cardiovascular Risk in Asymptomatic Adults: A Report of the American College of Cardiology. *J. Am. Coll. Cardiol.* 2010 56. e50-e103.

[36] Stamkopoulos, T; Diamantaras, K; Maglaveras, N; Strintzis, M. ECG analysis using nonlinear PCA neural networks for ischemia beat detection. *IEEE Trans. Signal Processing*. 1998 46. 3058-3067.

[37] Maglaveras, N; Stamkopoulos, T; Diamantaras, K; Pappas, C; Strintzis, M. ECG pattern recognition and classification using nonlinear transformations and neural networks: A review. *Int. J. Med. Inform.* 1998 52, 191-208.

[38] Maglaveras, N; Stamkopoulos, T; Pappas, C; Strintzis, M. ECG processing techniques based on neural networks and bidirectional associative memories. *J. Med. Eng. Technol.* 1998 22. 106-111.

[39] Papadimitriou, S; Mavroudi, S; Vladutu, L; Bezerianos, A. Ischemia detection with a self-organizing map supplemented by supervised learning. *IEEE Trans. Neural Networks*. 2001 12. 503-515.

[40] Papaloukas, C; Fotiadis, DI; Likas A; Michalis, LK. An ischemia detection method based on artificial neural networks. *Artificial Intelligence in Medicine*. 2002 24. 167-178.

[41] Tonekabonipour, H; Emam, A; Teshnelab, M; Shoorehdeli, MA. Ischemia prediction via ECG using MLP and RBF predictors with ANFIS classifiers. *ICNC*. 2011. 776-780.

[42] Gjertsson, P; Johansson, L; Lomsky, M; Ohlsson, M; Underwood, SR; Edenbrandt, L. Clinical data do not improve artificial neural network interpretation of myocardial perfusion scintigraphy. *Clin. Physiol. Funct. Imaging*. 2011 May 31(3). 240-245.

[43] Lindahl, D; Palmer, J; Ohlsson, M; Peterson, C; Lundin, A; Edenbrandt, L. Automated interpretation of myocardial SPECT perfusion images using artificial neural networks. *J. Nucl. Med.* 1997 Dec 38(12). 870-875.

[44] Stamkopoulos, T; Diamantaras, K; Maglaveras, N; Strintzis M. ECG analysis using nonlinear PCA neural networks for ischemia beat detection. *IEEE Trans. Signal Processing*. 1998 46. 3058-3067.

[45] Lindahl, D; Palmer, J; Edenbrandt, L. Myocardial SPET: artificial neural networks describe extent and severity of perfusion defects. *Clinical Physiology*. 1999 19(6). 497-503.

[46] Bagher-Ebadian, H; Soltanian-Zadeh, H; Setayeshi, S; Smith, ST. Neural network and fuzzy clustering approach for automatic diagnosis of coronary artery disease in nuclear medicine. *IEEE transactions on nuclear science*. 2004 51(1). 184-192.

[47] Papaloukas, C; Fotiadis, DI; Likas, A; Michalis, LK. An expert system for ischemia detection based on parametric modeling and artificial neural networks. *Proc. Eur. Med. Biol. Eng. Conf.* 2002. 742-743.

[48] Papaloukas, C; Fotiadis, DI; Likas, A; Michalis, LK. An ischemia detection method based on artificial neural networks. *Artificial Intelligence in Medicine*. 2002 24. 167-178.

[49] Lomsky, M; Gjertsson, P; Johansson, L; Richter, J; Ohlsson, M; Tout, D; van Aswegen, A; Underwood, SR; Edenbrandt, L. Evaluation of a decision support system for interpretation of myocardial perfusion gated SPECT. *Eur. J. Nucl. Med. Mol. Imaging*. 2008 Aug 35(8). 1523-1529.

[50] McMahon, EM; Korinek, J; Yoshifuku, S; Sengupta, PP; Manduca, A; Belohlavek, M. Classification of acute myocardial ischemia by artificial neural network using echocardiographic strain waveforms. *Comput. Biol. Med.* 2008 38(4). 416-424.

[51] Ture, M; Kurt, I; Kurum, AT; Ozdamar, K. Comparing classification techniques for predicting essential hypertension. *Expert Systems with Applications*. 2005 29, 583-588.

[52] Huang, S; Xu, Y; Yue, L; Wei, S; Liu, L; Gan, X; Zhou, S; Nie, S. Evaluating the risk of hypertension using an artificial neural network method in rural residents over the age of 35 years in a Chinese area. *Hypertens Res*. 2010 33. 722-6.

[53] Harrison, RF; Kennedy, RL. Artificial neural network models for prediction of acute coronary syndromes using clinical data from the time of presentation. *Ann. Emerg. Med.* 2005 46. 431-439.

[54] Lapuerta, P; Azen, SP; LaBree, L. Use of neural networks in predicting the risk of coronary artery disease. *Comput. Biomed. Res*. 1995 Feb 28(1). 38-52.

[55] George, J; Ahmed, A; Patnaik, M; Adler, Y; Levy, Y; Harats, D; Gilburd, B; Terrybery, J; Shen, GQ; Sagie, A; Herz, I; Snow, P; Brandt, J; Peter, J; Shoenfeld, Y.The prediction of coronary atherosclerosis employing artificial neural networks. *Clin Cardiol*. 2000 Jun 23(6). 453-456.

[56] Scott, JA; Aziz, K; Yasuda, T; Gewirtz, H. Integration of clinical and imaging data to predict the presence of coronary artery disease with the use of neural networks. *Coron Artery Dis*. 2004 Nov 15(7). 427-434.

[57] Srinivas, K; Raghavendra Rao, G; Govardhan, A. Analysis of Coronary Heart Disease and Prediction of Heart Attack in Coal Mining Regions Using Data Mining Techniques. *ICCSE 2010 The 5th International Conference on Computer Science and Education*. 1344-1349.

[58] Green, M; Björk, J; Forberg, J; Ekelund, U; Edenbrandt, L; Ohlsson, M. Comparison between neural networks and multiple logistic regression to predict acute coronary syndrome in the emergency room. *Artificial Intelligence in Medicine*, 2006 38, 305-318.

[59] Bulgiba, AM; Fisher, MH. Using neural networks and just nine patient-reportable factors of screen for AMI. *Health Informatics J.* 2006 Sep 12(3). 213-225.

[60] Allison, JS; Heo, J; Iskandrian, AE.Artificial neural network modeling of stress single-photon emission computed tomographic imaging for detecting extensive coronary artery disease. *Am. J. Cardiol.* 2005 Jan 95(2). 178-181.

[61] Atkov, OY; Gorokhova, SG; Sboev, AG; Generozov, EV; Muraseyeva, EV; Moroshkina, SY; Cherniy, NN. Coronary heart disease diagnosis by artificial neural networks including genetic polymorphisms and clinical parameters. *J. Cardiol.* 2012 Mar 59(2). 190-194.

[62] Chen, Q; Li, G; Leong, TY; Heng, CK.Predicting coronary artery disease with medical profile and gene polymorphisms data. *Stud. Health Technol. Inform.* 2007 129(Pt 2). 1219-1224.

[63] Haraldsson, H; Ohlsson, M; Edenbrandt, L.Value of exercise data for the interpretation of myocardial perfusion SPECT. *J. Nucl. Cardiol.* 2002 Mar-Apr 9(2). 169-173.

[64] Karabulut, EM; Ibrikçi, T. Effective diagnosis of coronary artery disease using the rotation forest ensemble method. *J. Med. Syst.* 2012 36. 3011–3018.

[65] Stefko K. Application of MLBP Neural Network for Exercise ECG Test Records Analysis in Coronary Artery Diagnosis. *Information Technologies in Biomedicine.* 2008. 179-183.

[66] Tham, CK; Heng, CK; Chin, WC. Predicting risk of coronary artery disease from dna microarray-based genotyping using neural networks and other statistical analysis tools. *J. Bioinform. Comput. Biol.* 2003 Oct 1(3). 521-539.

[67] Toft, J; Lindahl, D; Ohlsson, M; Palmer, J; Lundin, A; Edenbrandt, L; Hesse, B. The optimal reference population for cardiac normality in myocardial SPET in the detection of coronary artery stenoses: patients with normal coronary angiography or subjects with low likelihood of coronary artery disease? *Eur. J. Nucl. Med.* 2001 Jul 28(7). 831-835.

[68] Young, MT; Blanchard, SM; White, MW; Johnson, EE; Smith, WM; Ideker, RE. Using an artificial neural network to detect activations during ventricular fibrillation. *Comput. Biomed. Res.* 2000 33(1). 43-58.

[69] Chesnokov, YV. Complexity and spectral analysis of the heart rate variability dynamics for distant prediction of paroxysmal atrial fibrillation with artificial intelligence methods. *Artif. Intell Med.* 2008 43(2). 151-165.

[70] Martis, RJ; Krishnan, MM; Chakraborty, C; Pal, S; Sarkar, D; Mandana, KM; Ray, AK. Automated screening of arrhythmia using wavelet based machine learning techniques. *J. Med. Syst.* 2012 36(2). 677-688.

[71] Comak, E; Arslan, A; Türkoğlu, I. A decision support system based on support vector machines for diagnosis of the heart valve diseases. *Comput. Biol. Med.* 2007 37(1). 21-27.

[72] Avci, E. A new intelligent diagnosis system for the heart valve diseases by using genetic-SVM classifier. *Expert Systems with Applications*. 2009 36. 10618-10626.

[73] Nilsson, J; Ohlsson, M; Thulin, L; Höglund, P; Nashef, SA; Brandt, J. Risk factor identification and mortality prediction in cardiac surgery using artificial neural networks. *J. Thorac. Cardiovasc. Surg.* 2006 Jul 132(1). 12-19.

[74] Macrina, F; Puddu, PE; Sciangula, A; Trigilia, F; Totaro, M; Miraldi, F; Toscano, F; Cassese, M; Toscano, M. Artificial neural networks versus multiple logistic regression to predict 30-day mortality after operations for type a ascending aortic dissection. *Open Cardiovasc. Med. J.* 2009 Jul 7(3). 81-95.

[75] Ebell, MH. Artificial neural networks for predicting failure to survive following in-hospital cardiopulmonary resuscitation. *J. Fam. Pract.* 1993 36. 297-303.

[76] Bourdès, V; Ferrières, J; Amar, J; Amelineau, E; Bonnevay, S; Berlion, M; Danchin, N. Prediction of persistence of combined evidence-based cardiovascular medications in patients with acute coronary syndrome after hospital discharge using neural networks. *Med. Biol. Eng. Comput.* 2011 49(8). 947-955.

In: Network Topologies
Editor: Casandra Rendell

ISBN: 978-1-62618-180-9

Chapter 3

SIGNAL TRANSDUCTION PATHWAYS: WHEN NETWORK TOPOLOGY CAN EXPLAIN CELL FUNCTION

Nicola Bernabò, Mauro Mattioli and Barbara Barboni
Department of Comparative Biomedical Sciences,
University of Teramo, Teramo, Italy

ABSTRACT

The living beings continuously receive, process and respond to a myriad of signals. To this aim several transduction pathways have evolved, giving rise an amazingly complex functional dialogue into and among cells. Modern analytical techniques (such as 2D electrophoresis, DNA microarray, protein chips, etc...) increased dramatically our knowledge about the molecules involved in these signaling events and allowed the discovery of several biochemical determinants of signal transduction. Despite scientific community expectations, the applicative benefit of this always increasing information to date are scarce. In our opinion, this is due to the intrinsic complexity of biological beings, from molecules to cells, from tissues to organs, that make necessary the adoption of the computational modeling strategies. In this context, we adopted a biological networks-based approach to study the topology of signal transmission pathways. In particular we have realized and statistically analyzed the main topological indexes of biological networks representing the molecular events involved in 11 pathways of relevant importance for human biology [smooth and striated muscle contraction, release cycle of six neurotransmitters, visual phototransduction (rods), sperm capacitation, insulin signaling pathway, p53 pathway, regulation of retinoblastoma protein (pRb), mitochondrial ATP metabolism, glucose metabolism, signaling events mediated by stem cell factor receptor c-Kit and the circadian clock]. As a result we have found that the number of molecules composing each network is likely due to the balancing of energetic cost of each molecule and the need of system stability. In addition, we demonstrated that all the examined networks showed the same scale free topology and small world behavior with a clearly non hierarchical structure, unlike other biological networks. In particular the clustering coefficient (i.e. the measure of how each network node tend to cluster to other ones) was virtually zero and it was unrelated to the number of links per node, the characteristic path length (the measure of how many links it is necessary to pass through to travel between two random selected nodes) was about 6 and the averaged number of link per node ranged between 2 and 2.8. From these findings, it is possible to conclude that all examined pathways share the same topology that confers them some important biological features such as robustness against random failure, controllability, and specificity and efficiency in signal transmission, which are the specific signature of signal transduction pathways.

INTRODUCTION

The so called "technology-driven" research have made available for Scientists enormous amounts of data, thanks to the development and refinement of high- throughput technologies, such as 2D electrophoresis, DNA microarray, protein chips, mass spectrometry, etc… Contemporaneously, in latest years, biology has undergone a great revolution that culminated with the completion of Human Genome Project in 2003, leading to the sequencing of whole human genome (International Human Genome Sequencing Consortium, 2004). Before that date, it was beloved that this knowledge would immediately open immense opportunities for development and production of new drugs and diagnostics and therapeutics tools, with an immediate economic return. Despite these enormous expectation, the knowledge of the human genome and of various pathogenic agents genome sequences has led to the discovery of only a limited number of new drug targets (Drews, 2003). This, in one hand had dramatic market consequences, whilew on the other one, strengthen the concept that the greatest problem in modern biology is the data interpretation.

Indeed, biological systems are constituted by an ensemble of elements interacting each others in a non linear manner, giving rise to a network of interactions (Van Regenmortel, 2004). More in detail, they are characterized by well defined structural proprieties, typical of complex systems, such as:

- the elements composing the system are strongly interconnected, thus if one of them fails, possibly all the system fails;
- is often difficult to correctly define the links among the different components of the system, that are frequently non-linear and characterized by feed-back loops;
- biological systems are dynamical, thus evolve with time. As a consequence, prior states may have an influence on present states (the so called hysteresis). In other words, complex biological systems have a memory;
- the system components may themselves be complex systems. For instance, an organism is made up of tissues, which are made up of cells, which in turn, are made up of molecules. All of them are complex systems;
- complex biological systems, often, display emergent behaviors, i.e. they may show new properties when they are studied at a higher or lower levels. It is, for instance, the case of neuron networks or ant colonies. Each neuron or each ant *per se* obeys to a few simple rules, whereas a large number of them, reciprocally interacting is able to display a complex and sophisticated behavior (Anderson and McSea, 2001).

Consequently, it is impossible to study and analyze (biological) complex systems by using a reductionist approach, whereas it is necessary to consider them in their entirety (a complex system is more than the sum of its components) and adopting new strategies, different from those of traditional biology (Van Regenmortel, 2004). In particular, a promising approach is the computational modeling, which allows to realize models of complex events, thus allowing to infer important information.

One modeling strategy that offers great advantages is that based on networks theory. The concept that it is possible to describe and to analyze the reality as a network of interacting entities dates back more than 250 years ago, when in 1736 the Swiss mathematician and physicist Leonhard Euler solved the problem of Seven Bridges of Königsberg. The city of Königsberg in Prussia (now Kaliningrad, Russia) was set on both sides of the Pregel River, and included two islands connected to each other and to the mainland by seven bridges. The problem was to find a walk through the city, that would cross each bridge once and only once. The islands could not be reached by any route other than the bridges, and every bridge must have been crossed completely every time. Euler represented the land mass with a "vertex" or a node, and each bridge with a connection, an "edge" or a link. By this way, for the first time, the reality was represents by a mathematical model, i.e. by an abstract object: a network of nodes linked by edges representing their interactions (Barabási, 2003).

Now, we know that adopting this approach it is possible to study an amazing variety of natural and artificial phenomena, such as the market, the communications, the transports, the alimentary chains, the biochemical reaction that take place in cells, taking in account their complexity (Barabási, 2003). In this context the most important characteristic of networks is their architecture, the so called "network topology". The statistical analysis of networks topology, indeed, gives important quantitative, objective and comparable information, thus providing sophisticated inference about their function. More in detail it has been possible to define some different kinds of networks.

Random Networks

A random network (graph) is obtained by connecting a set of n nodes with random added edges. The most commonly studied model of random networks is that proposed by Edgar Gilbert, denoted $G(n,p)$, in which every possible edge occurs independently with probability p (Gilbert, 1959), and that proposed by Paul Erdős and Alfréd Rényi, denoted $G(n,M)$, in which it exists an equal probability of all graphs with exactly M edges (Erdős–Rényi or ER model) (Erdős and Rényi, 1959). Two different variants of ER model are known:

- the $G(n, M)$ model, in which a graph is chosen uniformly at random from the collection of all graphs which have n nodes and M edges. For example, in the $G(3, 2)$ model, each of the three possible graphs on three vertices and two edges are included with probability 1/3.
- the $G(n, p)$ model, in which a graph is realized by randomly connecting the nodes. In particular, each edge is added to the graph with a probability p, independent from every other edge. Thus, all

graphs with n nodes and M edges have equal probability of $p^M(1-p)^{\binom{n}{2}-M}$. The parameter p in this model represent a weighting function: as it increases from 0 to 1, the model becomes more and more likely to include graphs with more edges and less and less likely to include graphs with fewer edges. In particular, the case p = 0.5 corresponds to the case where all $2^{\binom{n}{2}}$ graphs on n vertices are chosen with equal probability.

Consequently, in random networks, the node degree (i.e. the number of links per node) follows a Poisson distribution, which indicates that most of nodes have approximately the same number of links, close to the average degree, that define the network scale. For instance, the water, at room temperature, is constituted by several cluster of molecules aggregated to form domains, due to the presence of hydrogen bonds. In particular, each water molecule binds on average 4 other molecules, thus it is possible to represent water as a network of scale = 4. The tail (high k region) of the degree distribution $P(k)$ decreases exponentially, implying that the nodes that significantly deviate from the average are extremely rare. The clustering coefficient is independent from the node degree, and the mean path length (i.e. the mean expected distance between two randomly selected connected nodes) is proportional to the logarithm of the network size, $l \sim \log N$ (Erdős and Rényi, 1959). From a biological point of view, random networks have two important features:

- the network behavior strictly depends on the network scale;
- random networks are democratic networks, i.e. each node has the same importance as others nodes in determining network proprieties.

Small Word Networks

Random networks are unable to explain a typical phenomenon of social and biological networks: the so called "small world" effect. In several experiment it was found that the navigability of real biological networks (i.e. the possibility to jump from a node to a other node) is greatest of that predicted by ER model. In particular the distance of two random selected nodes is shortest that expected. For instance, in classical Stanley Milgram experiment it was found that all the people are separated only from six step (thus originating the "six degrees of separation" myth) (Milgram, 1967). To take into account this phenomenon, in 1998 Duncan J. Watts and Steven Strogatz proposed a new model (known as Watts-Strogatz, WS, model)(Watts and Strogatz, 1998). WS model is the interpolation between an ER graph and a regular ring lattice: the underlying lattice structure of the model produces a locally clustered network, and the random links dramatically reduce the average path lengths.

Scale-Free Networks

WS model can explain important features of biological networks, which display a small world topology, as it happens, for instance, in neuronal networks, but are unable to describe another characteristic of biological (and non-biological) systems: the heterogeneity of node degree. Indeed, several phenomena, from the WWW connectivity to Internet architecture, from the stock market to the protein interaction networks, share the presence of a low number of most connected nodes (defined the "hubs") and a higher number of less connected nodes. In 1999 Albert Lazlo Barabási and Reka Albert proposed the so called Barabási–Albert (BA) model, which incorporates two important general concepts: growth and preferential attachment, thus originating scale-free networks (Barabási and Albert, 1999). In particular these networks are able to evolve, to grow, over the time, and the growth is realized by the preferential attachment of new nodes to the hubs. In other words, the more one node is connected, the more is its able to attract new connections to new nodes. New nodes are added to the network one at a time and each new node is connected to *m* already existing nodes with a probability that is proportional to the number of links that the existing nodes already have. Thus, the probability p_i that the new node is connected to node *I* is:

$$p_i = \frac{k_i}{\sum_j k_j}$$

where k_i is the degree of node *I* and the sum is made over all preexisting nodes *j* (Albert and Barabási, 2002). These networks are characterized by a power-law degree distribution: the probability that a node has *k* links follows $P(k) \sim k^{-\gamma}$, where γ is the degree exponent and usually ranges between 2 and 3. Then, the probability that a node is highly connected is statistically higher than in a random graph. As a consequence, the network behavior is strongly determined by a small number of highly connected nodes (the hubs) and a "typical" node does not exist (scale-free topology). These network do not have an inherent modularity, i.e. $C(k)$ is independent of k and have the average path length following l ~ log log N, which is significantly shorter than log N that characterizes random networks.

Hierarchical Networks

Hierarchical networks belong to the scale-free networks family but are characterized by modularity and high local clustering. In contrast to the BA scale-free models, where the clustering coefficient is independent of the degree of a specific node, in hierarchical networks the clustering coefficient can be expressed as a function of the degree in the following way:

$$C(k) \square k^{-\beta}$$

Several biological networks have been found to have a hierarchical topology, such as metabolic networks or protein interaction networks (Papin et al., 2004; Ravasz, 2009).

SIGNAL TRANSDUCTION

One of the cases in which biological modelization is enabling impressive scientific advances, is the study of biological data. Proof is that a new discipline, the systems biology, has been developed to bring together biology, physics, mathematics, engineering and informatics. Systems biology is aimed to collect, analyze and interpret biological data, viewing at the organism as an integrated network of interacting molecules: just these interactions are responsible for the organism form and functions.

An interesting application of systems biology, of continuously growing importance in the study of physiology and physiopathology, is the study of signal transduction. Indeed biological organisms are characterized by a multi-level organization, from atoms to molecules, from multi-molecular complexes to cells, tissues and organs. This implies the need of the development of mechanisms devoted to the coordination of different sub-systems. It is possible to affirm that the survival of living beings strongly depends on their ability to efficiently communicate and coordinate the messages arising from the different compartments and levels of organization, each of them characterized by the emergences of specific proprieties. As a consequence, a great number of molecules are devoted to the reception, processing and transfer of molecular messages, originating more or less complex signals transduction pathways, and specific structures, such as the nervous and the endocrine system. It is possible to schematize that every cell continuously receives and transfers a myriad of messages directed to itself (autocrine activity), to other near cells (paracrine activity) and to the whole organism (endocrine activity). All these signaling pathways involve hundreds or thousands of molecules interacting each other in several different manner, thus giving rise to an integrated cross-talking.

Despite several efforts have been conducted to describe and to study the organization of signaling systems, either in physiological or pathological conditions (Park et al., 2009; Vidal et al., 2011; Barabási et al., 2011; Kozuka-Hata et al., 2011; Rehemtulla, 2011), important aspect of their structure remain poorly understood. In particular it is not yet clear if the architecture of signal transduction systems, at cellular and tissue level, has a specific feature. To answer this intriguing question we examined 11 pathways of relevant importance in human biology, representing them as biological networks by means of a computational modeling-based approach. In particular were studied the smooth and striated muscle contraction, the neurotransmitter release cycle of six neurotransmitters [norepinephrine, acetylcoline, γ-aminobutyric acid (GABA), serotonin, glutamate and dopamine], the visual phototransduction (rods), the sperm capacitation, the insulin signalling pathway, the p53 pathway, the regulation of retinoblastoma protein (pRb), the mitochondrial ATP metabolism, the glucose metabolism, the signaling events mediated by stem cell factor receptor c-Kit (CD117) and the circadian clock.

Smooth and Striated Muscle Contraction

Muscle contraction is the process that leads to the contraction of rough and smooth muscle. In the first case, it makes possible the movement, the breathing, the social interaction, whereas the smooth muscle is involved in autonomous visceral homeostasis. Rough muscle contraction starts when acetylcholine diffuses across the synapse and binds to acetylcholine nicotinic receptors on the neuromuscular junction. Activation of the acetylcholine receptors leads to the entry of sodium into the cells and, as a consequence, the membrane depolarization. This electrical event promotes the activation of L-type voltage-dependent calcium channels in the T tubule membrane, which are close to calcium-release channels (ryanodine receptors) in the sarcoplasmic reticulum.

This causes the release of calcium from sarcoplasmic reticulum which, in turns, binds to troponin C present on the actin filaments of the myofibrils. The troponin, then, allosterically modulates the tropomyosin that moves, thus allowing the binding of myosin to the actin filament with contextual ATP hydrolysis. In particular, the release of ADP and P_i is coupled to the insurgence of power stroke.

In smooth muscle, the contraction starts with the depolarization of the sarcolemma followed by the extracellular calcium entering through L-type calcium channels. This first rise in intracellular calcium concentration causes a further release of the ion from the sarcoplasmic reticulum, mediated by ryanodine receptor channels and by inositol triphosphate receptor channels. The calcium binds calmodulin, which, in turn, activates myosin light-chain kinase. This enzyme phosphorylates myosin on the 20kDa myosin light chains, thus initiating contraction and activating the myosin ATPase.

The Neurotransmitter Release Cycle of Neurotransmitters (Norepinephrine, Acetylcoline, γ-Aminobutyric Acid (GABA), Serotonin, Glutamate and Dopamine)

In central nervous system (CNS), the neurons communicate each other by releasing molecular messages (the so called neurotransmitters), which can promote the activation or inactivation of target cells. These neurotransmitters are released in synaptic cleft as a result of the activation of a specific signal transduction pathway.

In particular neurotransmitters are synthesized in neurons and stored in vesicles in the axon terminal. When the onset of a action potential at this level determines the entry of extracellular calcium the vesicles are released in synaptic cleft.

Here, they diffuse and bind specific receptors on the post-synaptic terminal, thus determining an inhibitory (membrane hyperpolarization) or stimulatory (membrane depolarization) effect on post-synaptic neuron. Then, the neurotransmitters are removed by reuptake from the presynaptic terminal, or destroyed by specific enzymes.

The Visual Phototransduction (Rods)

The cells responsible for phototransduction are the rods and cones. Both these cells contain a 11-cis retinal, the aldehyde of Vitamin A1, bound to cell membrane protein opsin. Rods are active with low light level and do not mediate color vision, whereas cones can code the color comparing the outputs of the three different types of cones, each containing a opsin which respond to a different wavelength of light. In the dark the photoreceptor cells membrane is maintained in a depolarized status (with a membrane potential of about -40 mV, the so called "dark current") by high cGMP levels. This dark current keeps the cell depolarized at about -40 mV. The depolarization of the cell membrane promote the opening of voltage-gated calcium channels (VGCC), thus leading to an increase in intracellular calcium concentration. This event causes the fusion of neurotransmitter-containing vesicle with cell membrane and, consequently, the release of neurotransmitter in the synaptic cleft. The released neurotransmitter is glutamate, which determines the activation of postsynaptic neuron. In the light, when a photon is absorbed retinal undergoes isomerisation, changing from the 11-*cis* to all-*trans* configuration. This induces a conformational change of Opsin, to metarhodopsin II, thus activating the regulatory protein transducin. This causes transducin to dissociate from its bound GDP, and bind GTP, then the alpha subunit of transducin dissociates from the beta and gamma subunits, with the GTP still bound to the alpha subunit. The alpha subunit-GTP complex activates phosphodiesterase, that breaks down cGMP to 5'-GMP. The lower cGMP concentration causes the closure of sodium channels and, as a consequence, cell membrane hyperpolarization. As membrane hyperpolarizes, VGCC close, intracellular calcium concentration decreases and the amount of glutamate released by the cell also drops. A decrease in glutamate release determines the depolarization of On center bipolar cells (rod and cone On bipolar cells) and the hyperpolarization of off-center bipolar cells. Thus the stimulus is transduced to the CNS.

The Sperm Capacitation

Mammalian spermatozoa, once ejaculated, are not able to fertilize. Indeed these cells can recognize and bind the proteins of the oocyte glycoprotyeic coat, the zona pellucida (ZP), only after they have interacted with the female genital tract for a relatively long period, from hours to days, depending to the species. During this window of time, several physico-chemical modifications (collectively known as "capacitation") happen, involving the biochemical and biophysical asset of the whole sperm cell. The signaling pathways that drive the cellular response involves virtually all the biochemical aspect of spermatozoa metabolism: the intracellular calcium concentration rises (Abou-haila and Tulsiani DR, 2009) the protein phosphorylation pattern changes (Breitbart, 2002), the actin cytoskeleton reorganizes (Bernabò et al., 2011), the plasma membrane (PM) and the outer acrosome membrane (OAM) become more instable and gradually acquire the ability to fuse each other (Botto et al., 2010; Barboni et al., 2011), the spermatozoa motility is hyperactivated (Suarez, 2008). Once completed, the capacitation makes the spermatozoa able to recognize and to bind the zona pellucida of oocytes, thus promoting the onset of the acrosome content extrusion (acrosome reaction).

The Insulin Signaling Pathway

The insulin receptor (IR) is a transmembrane receptor activated by insulin, IGF-I, IGF-II, which belongs to the tyrosine kinase receptors class. IR plays a key role in the regulation of glucose homeostasis, a functional process that, under pathological conditions may result in a range of clinical manifestations including diabetes and cancer. The binding of ligand to the α-chains of the IR ectodomain induces a structural changes in the receptor itself, that leads to the autophosphorylation of various tyrosine residues localized within the intracellular TK domain of the β-chain. These changes promote the recruitment of specific adapter proteins such as the insulin receptor substrate proteins (IRS), Src, Src Homology 2-B (SH2-B), APS and proteins phosphatases including PTP1B, that activate the downstream pathways involved in glucose metabolism.

The Glucose Metabolism

As a result of the integrated function of hyper- and hypo-glycemic pathways the glucose levels are maintained in a physiological range. In particular the most important pathways are:

- glycolysis, in which the oxidation of glucose produce ATP and pyruvate;

- Krebs cycle, also known as the citric acid cycle, a series of chemical reactions in which energy is generated through the oxidization of acetate;
- the pentose phosphate pathway, in which hexoses are converted into pentoses and in NADPH;
- glycogenesis, in which the excess glucose is converted in glycogen,
- glycogenolysis - the breakdown of glycogen into glucose;
- gluconeoenesis - *de novo* synthesis of glucose molecules from other molecules, such as amino acids.

The p53 Pathway

p53 is a tumor suppressor protein that, in humans, is encoded by the *TP53* gene. It is involved in the regulation of the cell cycle and in preventing cancer, so that p53 has been defined "the guardian of the genome" (Read and Strachan, 1999). In normal cells p53 is inactivated by the binding with mdm2. Upon DNA damage or other stresses, it dissociate from the mdm2 complex and binds to the DNA, thus activating the expression of several genes including microRNA miR-34a, WAF1/CIP1 encoding for p21 and hundreds of other down-stream genes. As a result p53 will induce a cell cycle arrest to allow either repair and survival of the cell or apoptosis to destroy the damaged cell. What determine this choice and which are its molecular determinants is currently unknown (Suzuki and Matsubara, 2011).

The Regulation of Retinoblastoma Protein (pRb)

The retinoblastoma protein (abbreviated pRb, *RB* or *RB1*) is a tumor suppressor protein that belongs to the pocket protein family, that are characterized by a pocket for the functional binding of other proteins (Manning and Dyson, 2011). It pRb prevents the cell from replicating damaged DNA by preventing its progression along the cell cycle through G1 (first gap phase) into S (synthesis phase). Indeed pRb binds and inhibits transcription factors of the E2F family, which drive cells into S phase. The pRb-E2F/DP complex also attracts a histone deactylase (HDAC) protein to the chromatin, reducing transcription of S phase promoting factors, and suppressing DNA synthesis (Chinnam and Goodrich, 2011).

The Mitochondrial ATP Metabolism

In mitochondria of eukaryotes oxidative phosphorylation (OXPHOS) is a metabolic pathway that uses energy released by nutrients oxidation to produce adenosine triphosphate (ATP). During OXPHOS electrons are transferred from electron donors to electron acceptors (oxygen), in redox reactions, that release energy which is used to form ATP.

The energy released by electrons flows through an electron transport chain is used to transport protons across the inner mitochondrial membrane (chemiosmosis), thus generating potential energy in a pH gradient form and an electrical potential across this membrane. The enzyme ATP synthase uses this energy to generate ATP from ADP and P_i. This reaction is driven by the proton flow, which forces the rotation of a part of the enzyme.

The Signaling Events Mediated by Stem Cell Factor Receptor c-Kit (CD117)

Mast/stem cell growth factor receptor (SCFR), also known as proto-oncogene c-Kit or tyrosine-protein kinase Kit or CD117, is a protein that in humans is encoded by the KIT gene. CD117 is a cytokine receptor expressed on the surface of hematopoietic stem cells as well as other cell types, which binds the stem cells factor (SCF), also known as "steel factor " or “c-kit ligand” and which belong to the receptor tyrosine kinase type III classe.

When CD117 binds to the ligand, it forms a dimer that activates its intrinsic tyrosine kinase activity, that in turn phosphorylates and activates signal transduction molecules that propagate the signal in the cell. Signalling through CD117 plays a key role in cell survival, proliferation, and differentiation, as demonstrated by the finding that altered forms may be associated with the onset of cancer (Edling and Hallberg, 2007; Smalley et al., 2009).

Circadian Clock

A circadian rhythm is the biological process that displays an endogenous, entrainable oscillation of about 24 hours. Circadian rhythms are endogenous but could be influenced by external factors (the so called zeitgebers), such as the daylight. The classic circadian rhythm markers are (Klerman et al., 2002;

Benloucif et al., 2005) melatonin secretion by the pineal gland, core body temperature, plasma levels of cortisol.

Exploring Signal Transduction Topology

All the examined signaling pathways have been represented as biological networks and their topology has been statistically analyzed. To this aim, the data concerning human sperm capacitation were obtained from an already realized database (Microsoft Office Excel 2003) [9, 10], after updating, whereas the other datasets have been downloaded from Pathway Commons (http://www.pathwaycommons.org/pc/home.do)(Cerami et al., 2010) filtering the data for Homo sapiens, Taxonomy ID: 9606 at January 23-25, 2012. The networks were realized using the open access software Cytoscape 2.8.2 (http://www.cytoscape.org/) and their statistical and topological analyses have been carried out, considering the networks as undirected, by the Cytoscape plugin Network Analyzer. In particular we assessed the following topological parameters:

- the number of nodes: i.e. the total number of molecules involved;
- the number of edges: i.e. the total number of interaction found:
- the clustering coefficient: that is calculated as $CI = 2nI/k(k-1)$,where nI is the number of links connecting the kI neighbors of node I to each other. It is a measure of how the nodes tend to form clusters;
- the network diameter: i.e. the largest distance between two nodes;
- the averaged number of neighbors: i.e. the mean number of connection of each node;
- the characteristic path length: i.e. the expected distance between two connected nodes.

In addition, for each network, the node degree distribution law has been checked, and the exponent (γ) and the result of correlation test (r) have been calculated.

As a result we have found that the number of nodes and the number of edges ranged, respectively, between 125 and 802 and between 126 and 917. The clustering coefficient was comprised between 0 and 0.059 (median=0.002) while the diameter between 11 and 27 (median=15) and the averaged number of neighbors ranged between 2.016 and 2.798

(median=2.183). The characteristic path length ranged between 5.014 and 9.035 (median=6.453). See Table I. In addition it was found that all the networks shared a similar scale-free topology and that are characterized by a power law distribution of number of links per node with an exponent γ always ranging between -1.294 and -2.27 (median= -1.588).

Table 1. main topological indices of examined signal transduction pathways

	N° nodes	N° edges	Clustering coefficient	Diameter	Avg. n° neighbours	Char. path length	Nodes degree exponent (γ)	r
MC	134	139	0	12	2.075	5.590	-1.579	0.958
NRC	238	356	0.059	11	2.798	5.465	-1.294	0.977
VP	143	158	0	15	2.140	9.928	-2.015	0.964
SC	151	202	0.057	12	2.662	5.014	-1.588	0.817
INS	346	428	0.020	18	2.416	6.435	-1.428	0.993
p53	802	917	0.002	16	2.217	6.044	-1.464	0.907
pRb	291	323	0	15	2.151	6.814	-1.782	0.974
ATP	186	210	0.033	12	2.183	6.830	-1.532	0.983
GLU	260	346	0.025	17	2.723	6.502	-2.127	0.795
c-Kit	248	311	0	27	2.177	9.035	-1.799	0.677
CC	125	126	0	12	2.016	5.781	-1.590	0.978

The number of nodes represents the total number of molecules involved, the number of edges represents the total number of interaction found, the clustering coefficient is calculated as CI = 2nI/k(k–1),where nI is the number of links connecting the kI neighbors of node I to each other, the network diameter is the largest distance between two nodes, the Averaged n° neighbors represent the mean number of connection of each node, the Char. path length gives the expected distance between two connected nodes, exponent and r represents the exponent and the fitting result of power law equation of node distribution. MC=smooth and striated muscular contraction; NRC=6 neurotransmitters release cycle; VP=visual phototransduction (rods); SC=spermatozoa capacitation; INS=insulin *signaling pathway*; p53=p53 pathway; pRb=*regulation of* retinoblastoma protein; ATP=*mitochondrial* ATP metabolism; GLU=glucose metabolism; c-Kit=*signaling events mediated by stem cell factor receptor c-Kit*; CIRC=circadian clock.

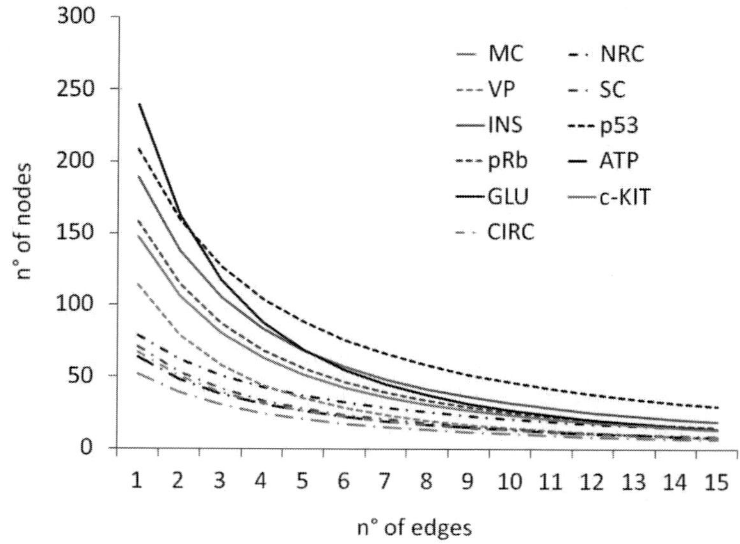

MC=smooth and striated muscular contraction; NRC=6 neurotransmitters release cycle; VP=visual phototransduction (rods); SC=spermatozoa capacitation; INS=insulin *signaling pathway*; p53=p53 pathway; pRb=*regulation of* retinoblastoma protein; ATP=*mitochondrial* ATP metabolism; GLU=glucose metabolism; c-Kit=*signaling events mediated by stem cell factor receptor c-Kit*; CIRC=circadian clock.

Figure 1. graph representing the node degree equation in examined pathways.

In all cases it was not found a correlation between the number of nodes and the topological parameters in all the examined networks ($p>0.05$), see Figure 2a, b, c, d, e.

Biological Inference

To explore the signaling cascades architecture, we have studied input (phototransduction) and output pathways (muscle contraction, sperm capacitation), information processing events (neurotransmitters release cycle), energy metabolism related pathways (ATP, glucose metabolism), endocrine control (insulin), cells cycle checkpoints (p53, pRb and c-Kit) and whole body coordination events (circadian clock), to examine the most various conditions as possible.

From the analysis of main topological indexes of examined networks, it was evident that the number of nodes (number of molecules considered) that make up the networks, ranges between 125 to 802, with a difference of almost one magnitude order. It is possible to hypothesize that this parameters could be conditioned by intrinsic factors.

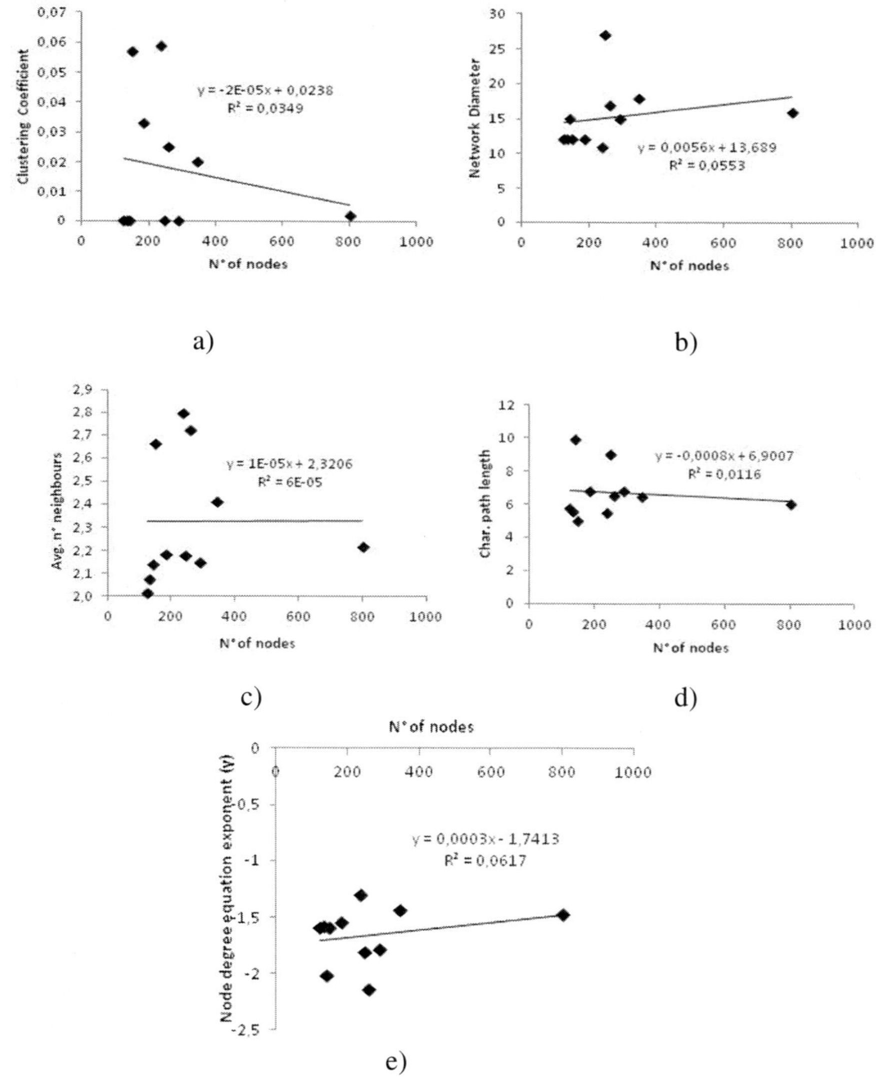

Figure 2. correlation analysis of topological indexes and number of nodes in examined networks.

In particular we suggest that the balance between energetic cost of each molecule and the need of system stability could determine the number of interacting molecules. In fact, it is conceivable that the metabolic investment to codify, to synthesize, to maintain, to modulate and, finally, to catabolize each molecule may have determined the transduction pathways evolution characterized by the smaller number of molecules, with the aim to minimize energy expenditure.

At the same time, the need to ensure a high system stability could favor the evolution of pathways composed by the largest number of molecules: indeed any fluctuation, or failure, of each element will affect in a much more limited way the whole system the higher is the number of its components.

In addition, the networks topological indexes analysis makes immediately evident that all networks exhibit the same topological feature: they all are scale free networks (see Table 1 and Figure 2), accomplishing to the BA model. Indeed, the exponents of node degree distribution were very similar, ranging from -2.127 to -1.294 (median -1.588), thus conferring to all the examined networks a similar connectivity (see Figure 1). In the classical BA model γ ranges from 2 to 3 (Albertand Barabási, 2002), whereas in our networks, it is comprised between 2 and 3 in less than 20% of cases (2/11) while it is lower in the other cases.

This relief, in our opinion, could be due to two different factors:

- the relatively small amount of molecules involved;
- the relatively high amount of highly connected ones.

This is not in contrast with BA model but it represents a specific feature of this general model, as it happens in a myriad of biological and, mostly, non biological entities.

The scale free topology could have important consequences on networks biological proprieties conferring it:

- *the robustness against random failure*. The robustness is guarantee by the condition that a random perturbation, in most of cases, will involve the most frequent typology of nodes (i.e. the less connected ones) thus having a negligible consequence on network architecture, and, as a consequence, on whole cellular function (Albert et al., 2000; Albert, 2005). In our model, we can estimate that exist only a probability <10% that one of them will be involved in random damages.

- *the controllability*. The small number of highly linked nodes implies that the whole system can be efficiently modulated by acting on a small number of molecules, thus reducing the energetic cost and facilitating/accelerating the cell response. In comparison, if the same networks had a "democratic" structure with all the nodes having the same number of links and the same importance in determining network topology, the cell behavior would be under the control of tens or hundreds of different molecules (i.e. 1 to 2 order of magnitude more than in scale free networks). This latter condition would require an enormous increase in system complexity and a dramatically higher energetic cost to coordinate and control the different signaling pathways. Incidentally, this topology could represent an advantage when an external manipulation of biological events is carried, as it happens in case of technological or pharmacological applications. Ultimately, the identification of a limited numbers of hubs could be helpful in controlling the biological systems.
- the easy navigability. All networks displayed an average numbers of link per node ranging from 2.1 to 2.7 and almost half of the nodes (~40%) were connected exactly by two links: one in input and one in output. This finding allows us to suppose that the networks are able to quickly transfer messages in a specific manner, avoiding redundancy or loops. In keeping with this hypothesis, half of the networks (55%) were characterized by a very low value of clustering coefficient (less 0.05) and in the remaining (45%; 5/11) is 0, thus suggesting to exclude the presence of hierarchical patterns. This, in our opinion, is a very important finding. Indeed, it is generally though that networks architecture representing biological phenomena is highly hierarchical and clustered (Wuchty et al., 2003, Papin et al., 2004; Ravasz, 2009). This is likely true for large scale genetic networks, where it exist a well defined modularity of nodes subsetand a hierarchy in genes expression. On the contrary, signaling networks must respond, in the shorter possible time and in the most specific possible way, to variation in cell homeostasis. As a consequence, a different organization could be evolved, in which the molecular message is carried, directionally, from the beginning to the end of the chain, avoiding the presence of factors (such as loops or clusters) that could slow or interfere with the propagation of messages themselves. In this context, as showed in Table 1, the characteristic path length ranges from 5.7 to 6.7 except that for visual transduction and c-kit pathway

> that is ~10. A short path length may have two important physiological consequences on signaling cascade. As first, if within a signaling pathway each molecule can interact with the others through a small number of passages, the loss of information due to the signal decrease will be minimized and, consequently, the signaling efficiency will be maximized. Secondly, any local signaling perturbation could influence the whole network in a short time, thus increasing the system responsiveness to intracellular and extracellular stimuli. This condition is in agreement with the classical small word concept, originally observed in social studies and subsequently found in several other systems, from neural networks to the World Wide Web. In this context, it is important to note that, as demonstrated by Latora and Marchirori (Latora and Marchirori, 2001), the real data analysis indicates that the networks characterized by the small-world behavior are both globally and locally efficient.

It is interesting, also, to note that all the examined topological indexes were statistically unrelated (Figure 2) to the number of nodes (in all the cases $p>0.05$), thus suggesting that the networks sizedid not influence its proprieties, and then, once again a scale free behavior.

These data are further supported by our recent findings demonstrating that biological networks of signaling transduction pathways involved in spermatozoa activation of organisms with different reproductive physiology and ecology (sea urchin, Caenorhabtidis elegans and humans), share a similar topology (Bernabò et al., 2010; Bernabò et al., 2011; Bernabò et al., 2012a; Bernabò et al., 2012b).

CONCLUSION

In conclusion, taken alltogether, these findings seem to suggest that signaling networks have a specific, well established topology, different from others biological networks, such as protein interaction networks or genes expression networks. In particular:

- the number of nodes is the balancing resultof each molecule energetic costand the need of system stability;
- the signaling networks are easily navigable and controllable;

- the scale free architecture and the small world behavior coexist with the absence of clustering;
- the specificity and the fastness of signals transfer, seem to represent the signature of these networks.

It is conceivable that this architecture could be a universal feature of biological messages transduction systems, probably due to its evolutionary advantages.

REFERENCES

Abou-haila, A., Tulsiani, D. R. Signal transduction pathways that regulate sperm capacitation and the acrosome reaction. *Arch. Biochem. Biophys.* 2009 485, 72-81.

Albert, R. Scale-free networks in cell biology. *J. Cell Sci.* 2005 118: 4947-4957.

Albert, R., Barabási, A. L. Statistical mechanics of complex networks. *Rev. Mod. Phys.*2002 74: 47-97.

Albert, R., Jeong, H, Barabási, A. L. Error and attack tolerance of complex networks. *Nature* 2000 406: 378-382.

Anderson, C., McShea, D. W. Individual versus social complexity, with particular reference to ant colonies. *Biol. Rev. Camb. Philos. Soc.* 2001 76: 211-237.

Barabási, A. L. *Linked: How Everything is Connected to Everything Else and What It Means for Business, Science, and Everyday Life.* 2003 New York: Plume.

Barabási, A. L., Albert, R. Emergence of scaling in random networks. *Science.* 1999 286: 509–512.

Barabási, A. L., Gulbahce, N., Loscalzo, J. Network medicine: a network-based approach to human disease. *Nat. Rev. Genet.* 2011, 12: 56-68.

Barabási, A. L., Oltvai, Z. N. Network biology: understanding the cell's functional organization. *Nature Reviews Genetics 2004* 5, 101-113.

Barboni, B., Bernabo`, N., Palestini, P., Botto, L., Pistilli, M. G., Chiarini, M., Tettamanti, E., Battista N., Maccarrone, M., Mattioli M. Type-1 Cannabinoid Receptors Reduce Membrane Fluidity of Capacitated Boar Sperm by Impairing Their Activation by Bicarbonate. *PLoS ONE* 2011 6, e23038.

Benloucif, S., Guico, M. J., Reid, K. J., Wolfe, L. F., L'hermite-Balériaux, M., Zee, P. C. Stability of melatonin and temperature as circadian phase markers and their relation to sleep times in humans. *Journal of Biological Rhythms* 2005 20: 178–188.

Bernabò, N., Berardinelli, P., Mauro, A., Russo, V., Lucidi, P., Mattioli, M., Barboni, B. The role of actin in capacitation-related signaling: an in silico and in vitro study. *BMC Systems Biology,* 2011 30: 47.

Bernabò, N., Mattioli, M., Barboni, B. A Systems Biology Approach to Understanding Male Infertility. In: *Male Infertility*, Edited by Anu Bashamboo and Kenneth David McElreavey, Published by InTech, 2012a.

Bernabò, N., Mattioli, M., Barboni, B. The spermatozoa caught in the net: the biological networks to study the male gametes post-ejaculatory life. *BMC Systems Biology*, 2010 4, 87.

Bernabò, N., Saponaro, I., Mattioli, M., Barboni, B. Signaling Strategy in Spermatozoa Activation in Sea Urchin, C. elegans and Human: Three Different Players for the Same Melody. *J. Bioengineer. and Biomedical Sci.* 2012b 001 (Suppl. 3).

Botto, L., Bernabò, N., Palestini, P., Barboni, B. Bicarbonate induces membrane reorganization and CBR1 and TRPV1 endocannabinoid receptor migration in lipid microdomains in capacitating boar spermatozoa. *Journal of Membrane Biology,* 2010 238: 33-41.

Breitbart, H. Intracellular calcium regulation in sperm capacitation and acrosomal reaction. *Mol. Cell Endocrinol.*, 2002 187: 139 - 144.

Cerami, E. G., Gross, E. B., Demir, E., Rodchenkov, I., Babur, Ö., Anwar, N., Schultz, N., Bader, G. D., Sander, C. Pathway Commons, a web resource for biological pathway data. *Nucl. Acids Res.* 2012 39: 685-690.

Chinnam, M., Goodrich, D. W. RB1, development, and cancer. *Curr. Top. Dev. Biol.* 2010 94: 129-169.

Drews, J. Stategic trends in the drug industry. *Drug Discov. Today* 2003 8: 411-420.

Edling, C. E., Hallberg, B. c-Kit--a hematopoietic cell essential receptor tyrosine kinase. *Int. J. Biochem. Cell Biol.* 2007 39: 1995–1998.

Erdős, P., Rényi, A. On Random Graphs I. *Publ. Math. Debrecen* 1959 6: 290–297.

Gilbert, E. N. Random graphs. *Annals of Mathematical Statistics* 1959 30: 1141–1144.

International Human Genome Sequencing Consortium. Finishing the euchromatic sequence of the human genome. *Nature* 2004 431: 931-945.

Klerman, E. B., Hayley, B., Gershengorn, J. F., Duffy, Richard., Kronauer, E. Comparisons of the Variability of Three Markers of the Human Circadian Pacemaker. *J. Biol. Rhythms* 2002 17: 181–193.

Kozuka-Hata, H., Tasaki, S., Oyama, M. Phosphoproteomics-based systems analysis of signal transduction networks. *Front Physiol.* 2011 2:113.

Latora, V., Marchiori, M. Efficient Behavior of Small-World Networks. *Phys. Rev. Lett.* 87: 2001 198701-198711.

Manning, A. L., Dyson, N. J. pRB, a tumor suppressor with a stabilizing presence. *Trends Cell. Biol.* 2011 21: 433-441.

Milgram, S. The Small World Problem. *Psychology Today* 1967 2: 60–67.

Papin, J. A., Reed, J. L., Palsson, B. O. Hierarchical thinking in network biology: the unbiased modularization of biochemical networks. *Trends Biochem. Sci.* 2004 29:641-647.

Park, J., Lee, D. S., Christakis, N. A., Barabási, A. L. The impact of cellular networks on disease comorbidity. *Mol. Syst. Biol.* 2009 5: 262.

Ravasz, E. Detecting hierarchical modularity in biological networks. *Methods Mol. Biol.* 2009 541:145-160.

Ravasz, E. B., Barabási, A. L. Hierarchical organization in complex networks. *Physical Review E* 2003 67.

Read, A. P., Strachan, T. Chapter 18: Cancer Genetics. *Human molecular genetics* 2. 1998 New York: Wiley.

Rehemtulla, A. The interconnectedness of cancer cell signaling. *Neoplasia.* 2011 13:1183-1193.

Smalley, K. S., Sondak, V. K., Weber, J. S. c-KIT signaling as the driving oncogenic event in sub-groups of melanomas. *Histol. Histopathol.* 2009 24: 643-650.

Suarez, S. S. Control of hyperactivation in sperm. *Hum. Reprod. Update* 2008 14: 647-657.

Suzuki, K., Matsubara, H. Recent advances in p53 research and cancer treatment. *J. Biomed. Biotechnol.* 2011 978312.

Van Regenmortel, M. H. V. Reductionism and complexity in molecular biology. *EMBO reports* 2004 11: 1016-1020.

Vidal, M., Cusick, M. E., Barabási, A. L. Interactome networks and human disease. *Cell* 2011, 144: 986-98.

Watts, D., Strogatz, S. H. Collective dynamics of 'small-world' networks. *Nature* 393 1998 4: 09–10.

Wuchty, S., Oltvai, Z. N., Barabási, A. L. Evolutionary conservation of motif constituents in the yeast protein interaction network. *Nat. Genet.* 2003 35:176-179.

In: Network Topologies
Editor: Casandra Rendell
ISBN: 978-1-62618-180-9

Chapter 4

MODEL-BASED RISK ANALYSIS OF COMPLEX NETWORKS

Ted G. Lewis[1*]
[1]Center for Homeland Defense and Security
Naval Postgraduate School, Monterey, CA, US

ABSTRACT

Model-based Risk Analysis (MBRA) is a method and tool for analyzing complex system risk as defined by a network of components and their links to one another. Network risk is defined as the expected consequence across all nodes and links of the network. Network nodes and links are the fundamental assets represented in MBRA. Nodes can be anything – a building, tunnel, computer, water pump, reservoir, power substation, person, group, or telecommunication switching office. Links can be anything: a power line, familial relation, pipe, Internet fiber optic cable, or virtual dependency. Nodes and links have properties: values needed to compute risk as well as network properties such as number of connections, betweeness, and height.

MBRA employs fault trees to model multiple threats arrayed against single or multiple nodes; and expected utility theory (EUT) to compute overall network risk. In addition to calculating EUT risk, MBRA performs various modeling and simulation functions as follows:

* Corresponding author: Ted G. Lewis. Center for Homeland Defense and Security, Naval Postgraduate School, Monterey, CA. 93943, E-mail: tlewis@nps.edu.

1 Network properties: compute degree, betweeness, height, and contagiousness.
2 Objectives: optimize on risk, vulnerability, probability of failure, and consequence.
3 Resource allocation: distribute budgets across nodes and links such that one of the objectives listed in 2 is optimized.
4 Return on Investment Analysis: analyze changes in objectives versus investment in threat, prevention, and response capability.
5 Cascades: simulate cascade failures precipitated by a single failure.
6 Flow: simulate the loss of flows through the network due to a single failure.
7 Displays: calculate and display the exceedence probability generated by simulation of faults in the cascade and flow simulations.
8 Threat analysis: model threats versus nodes and links as a fault tree.
9 Game theory: simulate a 2-person game and find optimal allocations of threat, vulnerability, and response investments using Stackleberg optimization.
10 Exceedence probability analysis and probable maximum loss analysis (an approach favored by the insurance industry).

MBRA has been used since 2006 to analyze dozens of critical infrastructure systems of interest to homeland security. Applications have mainly come from students doing classroom projects. Experience has showed that MBRA can be extremely expressive and capable of analyzing complete systems of all kinds and sizes. Some of MBRA's limitations have become apparent: threat, vulnerability, and consequence decrease as an exponential function of investment – a model that may not represent reality. Additionally, cascade failure probability in some systems, such as a power grid collapse or flood, actually increases as a collapse is taking place. MBRA assumes likelihood of collapse is a constant. Finally, MBRA does not model dynamical systems with feedback loops. It assumes directional or bidirectional flows, but does not model dynamic feedback. MBRA is freely available at www.CHDS.us for Windows and Macintosh operating systems.

INTRODUCTION

Risk analysis has a long and rich history rooted in the original work of Daniel Bernoulli, the father of expected utility theory, EUT (Bernoulli 1954). Building on the pioneering work of the inventor of modern probability theory Blaise Pascal, Bernoulli defined probabilistic risk as the expected outcome or consequence of n events each occurring with probability $Pr(e_i)$ and incurring a cost of $c(e_i)$:

$$Z = \sum_{i=1}^{n} \Pr(e_i)c(e_i)$$

This definition is simply the expected value of a series of events weighted by the likelihood of their occurrence. In Bernoulli's words, "Expected values are computed by multiplying each possible gain by the number of ways in which it can occur, and then dividing the sum of these products by the total number of possible cases where, in this theory, the consideration of cases which are all of the same probability is insisted upon". Bernoulli equated expected value, utility, and risk with Z, explaining, "...and the profit which corresponds to this utility will equal the value of the risk in question". Risk analysis and expected utility theory (EUT) began in 1738 with Bernoulli's seminal paper.

In the 20th century, EUT and risk analysis were fundamental to many theories in economics, game theory, and decision theory. For example, the famous von Neumann-Morgenstern axioms used to model rational actors in game theory are based on EUT. One of the most controversial and lasting applications of EUT is in the formulation of a method of risk analysis called PRA – Probabilistic Risk Analysis – proposed by Norman Rasmussen of Massachusetts Institute of Technology.

Rasmussen applied EUT to the problem of estimating the risk to human life posed by nuclear power plants (Rasmussen 1975). He extended EUT to a new level of analysis by introducing fault trees – a combination of EUT and logic that models the interaction of the component parts of a system such as a power plant. PRA combines AND, OR, XOR logic, EUT, and De Morgan's Law for computing the likelihood of coupled events: event A AND B; A OR B; A XOR B, and so on. MBRA extends fault tree analysis even further by adding resource allocation and risk minimization to the basic PRA method.

Probabilistic risk analysis, PRA, defines risk z as the product of vulnerability, v, and consequence, c: z = v c, where v is the probability that a component or asset will fail if attacked or stressed, and c is the (financial or fatality) consequence of a failure or fault. In some fields, z is associated with a hazard such as a hurricane or attack by a terrorist. Hazards occur with a certain *a priori* or *a posteriori* probability, and incur a certain damage or consequence.

In most applications, vulnerability is an *a priori* estimate of the probability of failure, and consequence is typically measured in dollars or lives. Rasmussen's PRA extended EUT by modeling a system or series of events as a tree with nodes representing a system component, leaves representing hazards, and logic gates representing the relationships among components and hazards. For example, a redundant design might incorporate two identical components A and B. Failure of the system cannot occur unless both A and B fail. This is modeled in a fault tree as A AND B, with corresponding likelihood of Pr(A)Pr(B). Alternatively, a system may fail if either A or B or both A and B fail, in which case the likelihood of failure is expressed as the probability of (A OR B) occurring. We can use De Morgan's Law to calculate the likelihood of A, B, or A and B failing: Pr(A OR B) = 1 – [(1-Pr(A)(1-Pr(B)]. Fault trees can be hierarchically organized so that levels are connected together through application of AND, OR, and XOR connectors.

PRA has been criticized for lacking in expressive power. Principally, PRA and its main tool, fault trees, cannot model elaborate interactions among complex adaptive systems such as the propagation of a simple fault in a power grid, or interdiction of flow through a gas or oil pipeline system with alternative paths and partial failure modes. That is, fault trees lack the expressiveness to model cascade failures (domino effects) common in power outages (Final Report 2003), structure and vulnerability of social networks such as a terrorist network (Arquilla 2005), and sequences of faults occurring in normal accident theory (Perrow 1999). Furthermore, fault trees do not model dynamic properties of systems such as the flow of a commodity through a telecommunications network, water, gas or oil pipeline system, or the spread of a contagion through a human population.

To overcome the expressiveness limitations of fault trees, modelers have turned to network science as a tool to model complex, dynamic, and highly interconnected systems. Barabasi, Buchanan, Albert, and others pioneered the application of network science to complex systems analysis (Albert 2000, Strogatz 2001, Barabasi 2002, Buchanan 2002, and Watts 2003). These early pioneers focused on the topological structure of complex systems modeled as abstract graphs or networks. However, they ignored the potential of combining network science with PRA as a foundation for risk analysis of complex systems.

Lewis combined network science with PRA to define network risk – an adaptation of PRA to model highly connected and interdependent systems (Lewis 2006). The result was a model-based risk analysis method and tool, MBRA, for analyzing systems of all sizes, regardless of their complexity, using the abstract nature of networks. Network modeling is perhaps the simplest possible method of representation embracing nearly any system with components and relationships among the components. Nodes can represent anything, and links can represent physical or relational connections among nodes. The power of this abstraction enables modelers to represent almost any system as a network with PRA-like parameters. MBRA extends PRA in an easy-to-apply framework with greater expressive power than PRA. [However, MBRA too has limitations].

MBRA is a method and tool for analyzing complex system risk as defined by a network of components and their links to one another. Network risk is defined as the expected consequence across all nodes and links of the network. Network nodes and links are the fundamental assets represented in MBRA. Nodes can be anything – a building, tunnel, computer, water pump, reservoir, power substation, person, group, or telecommunication switching office. Links can be anything: a power line, familial relation, pipe, Internet fiber optic cable, or virtual dependency. Nodes and links have properties: values needed to compute risk as well as network properties such as number of connections, betweeness, and height.

MBRA uses fault trees to model multiple threats arrayed against single or multiple nodes; and EUT to compute overall network risk. In addition to calculating EUT risk, MBRA performs various modeling and simulation functions as follows:

1 Network properties: compute degree, betweeness, height, and contagiousness.
2 Objectives: optimize on risk, vulnerability, probability of failure, and consequence.
3 Resource allocation: distribute budgets across nodes and links such that one of the objectives listed in 2 is optimized.
4 Return on Investment Analysis: analyze changes in objectives versus investment in threat, prevention, and response capability.
5 Cascades: simulate cascade failures precipitated by a single failure.
6 Flow: simulate the loss of flows through the network due to a single failure.
7 Displays: calculate and display the exceedence probability generated by simulation of faults in the cascade and flow simulations.
8 Threat analysis: model threats versus nodes and links as a fault tree.

9 Game theory: simulate a 2-person game and find optimal allocations of threat, vulnerability, and response investments using Stackleberg optimization.
10 Exceedence probability analysis and probable maximum loss analysis (an approach favored by the insurance industry).

The number of inputs to MBRA were purposely kept to a minimum to facilitate ease of use. Every node and link requires an estimate of threat probability, vulnerability probability, consequence, prevention cost, and response cost. These five node and link attributes plus structural properties obtained from the network, itself, are sufficient to compute risk and other results as described in the list above. MBRA has been used since 2006 to analyze dozens of critical infrastructure systems of interest to homeland security. Applications have mainly come from students doing classroom projects. Experience has showed that MBRA can be extremely expressive and capable of analyzing complete systems of all kinds and sizes. Some of MBRA's limitations have become apparent: threat, vulnerability, and consequence decrease as an exponential function of investment – a model that may not represent reality. Additionally, cascade failure probability in some systems, such as a power grid collapse or flood, actually increases as a collapse is taking place. MBRA assumes likelihood of collapse is a constant.

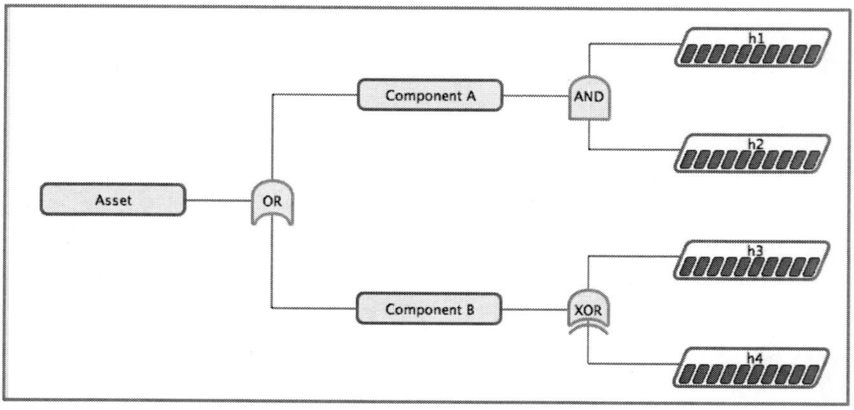

Figure 1(a). Fault tree of an asset with two components A and B. Component A has two hazards, h_1 and h_2, and component B has two hazards h_3 and h_4. Component A fails only if h_1 and h_2 fail; component B fails if either h_3 or h_4 fails, but not both. The asset fails if component A or B or both A and B fails. AND gates mean "both" must fail; OR gates mean one or more must fail; and XOR means only one, but not more than one, must fail for the component to fail.

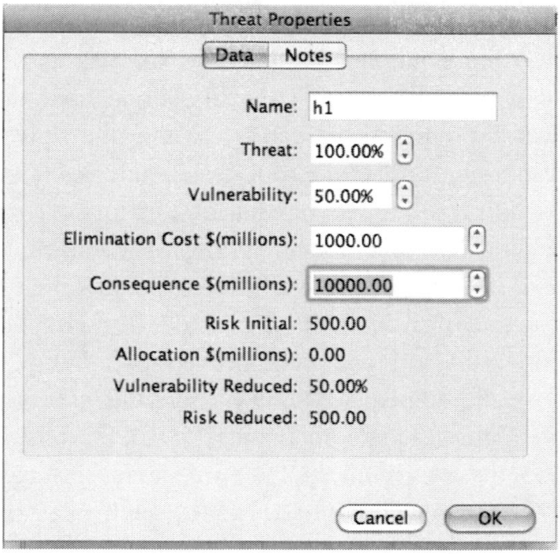

Figure 1(b). Data input dialog for fault tree hazards. User inputs are the name, threat and vulnerability probabilities as percentages, and consequence as a dollar amount. Threat t_i, is the probability of the hazard occurring; vulnerability $v_i(0)$ is the initial likelihood of component failure due to the hazard; elimination cost is the estimated cost to reduce vulnerability to a minimum, $V_i(\infty)$, typically 1-5% of the initial vulnerability; and consequence c_i is the damage caused by the hazard. The user enters these values. Initial risk, optimal allocation of a fixed budget to reduce the vulnerability of this hazard, reduced vulnerability after the optimal allocation, and reduced risk are computed by MBRA.

Finally, MBRA does not model dynamical systems with feedback loops. It assumes directional or bidirectional flows, but does not model dynamic feedback. MBRA is freely available at www.CHDS.us for Windows and Macintosh operating systems.

FAULT TREE ANALYSIS

PRA is often equated with fault tree analysis, whereby an asset under study is decomposed into components subject to a variety of hazards: h_1, h_2, h_3,... h_k; each hazard is paired with an asset or component and is assumed to occur with threat probability t_1, t_2, t_3,... t_k.

The component-hazard pairs of interest are placed into a fault tree as shown in Figure 1, connected by logic relations, AND, OR, or XOR.

Thus a fault tree is a logic model of the likelihood of asset failure when one or more of its components fail. In this case, the risk associated with each component-hazard pair, $z_i = t_i v_i c_i$ and the likelihood of the entire fault tree failing is a Boolean expression obtained by tracing the flow of one or more faults from one or more hazards to the root of the fault tree (Lewis, 2006).

The PRA and fault tree models have a long history of use in the nuclear power industry and more recently in MSRAM (Maritime Security Risk Assessment Method) developed by the US Coast Guard.

It is a capable and proven method of modeling simple risk when the number of component-hazard pairs are limited to a few dozen, and it is reasonable to assume that logical AND, OR, and XOR gates accurately model the dynamic relationships among components.

MBRA implements an enhanced fault tree analysis algorithm with the addition of two major innovations. First, MBRA fault trees relate investment to reduction in vulnerability. Operators can buy-down vulnerability and therefore risk.

Vulnerability is taken to mean the probability of total destruction of the asset or component of the asset should the designated hazard occur or attack be attempted. In MBRA vulnerability is modeled as an exponential function, $v(p_i)$, that monotonically declines from its initial input value, $v_i(0)$, to a minimum value, $v_i(\infty)$, as an investment is made to protect an asset or component.

Vulnerability is never zero, however, because an infinite investment would be required to entirely erase vulnerability.

Figure 1(b) shows the dialog for inputting these parameters into one component-hazard pair at a time, and Figure 2 illustrates calibration of exponential functions used in MBRA.

A second innovation implemented by MBRA is an algorithm to optimally allocate a budget to hazards such that overall risk is minimized. Given a budget P and a fault tree containing hazards as defined by threat, initial vulnerability, consequence, and vulnerability elimination cost, $p_i(\infty)$, MBRA allocates P to each hazard such that total fault tree risk is minimized. MBRA defines risk in fault trees as the total expected loss from all hazards:

$$R = \sum_{i}^{n} t_i\, v(\mathrm{p_i})\, c_i$$

$$v(\mathrm{p_i}) = v_i(0)e^{-\gamma_i \mathrm{p_i}}; \quad \gamma_i = -\ln[\frac{v_i(\infty)}{v_i(0)}]/\mathrm{p_i}(\infty)$$

For each component or component-hazard pair i:

- t_i: probability that hazard or attack occurs, threat input in Figure 1.
- $v(p_i)$: model of the probability of destruction vice investment, if hazard occurs
- p_i: investment to reduce vulnerability from initial value of $v_i(0)$, calculated
- $\mathrm{p_i}(\infty)$: investment that reduces vulnerability to $\mathrm{v_i}(\infty)$, elimination cost in Figure 1.
- $\mathrm{v_i}(\infty)$: vulnerability after an investment of $\mathrm{p_i}(\infty)$, set parameter typically .01.
- $v_i(0)$: initial vulnerability of asset or component, vulnerability in Figure 1.
- c_i: consequence of a successful attack on asset or component, Figure 1.
- n: number of hazards in fault tree.

The hazard data is constrained by a budget, P:

$$P = \sum_{i}^{n+m} p_i;$$

$p_i \geq 0$: P is total budget available to reduce vulnerability, given Fault tree vulnerability, or probability that the entire fault tree fails, is obtained by working up the fault tree from hazards at the lowest level to the root of the fault tree while applying the following equations for propagated vulnerability. Hazards connected to a component by an AND gate produce a failure in the component with probability defined by the product of hazard vulnerability values; OR gates with probability defined by De Morgan's Law; and XOR gates with probability defined by the sum of exclusive failure event probabilities. The equations in each case are:

AND gate

$$\Pr(AND) = \prod_{s \in component} t_s\, v(p_s);$$

component is the set of hazards or components attached to *AND*

OR gate

$$\Pr(OR) = 1 - \prod_{s \in component} [1 - t_s\, v(p_s)];$$

component is the set of hazards or components attached to *OR*

XOR gate

$$\Pr(XOR) = \sum_{s \in component} \left\{ t_s\, v(p_s) \prod_{j \neq s} [1 - t_j\, v(p_j)] \right\};$$

component is the set of hazards or components attached to *XOR*

For example, the fault tree in Figure 1 is reduced to a single expression for vulnerability by applying these equations from the lowest hazards up to the fault tree root. Let the initial product of threat and vulnerability, $t_i v_i(0)$, corresponding with hazards h_1, h_2, h_3, and h_4 be 80%, 50%, 40%, and 20%, respectively. The corresponding initial probabilities of component failure and fault tree vulnerability are calculated as follows.

Hazards h1 AND h2

$$\Pr(AND) = \prod_{s=1}^{2} t_s\, v_s(0) \;=\; (.80)(.50) \;=\; .40$$

Hazards h3 XOR h4

$$\Pr(XOR) = \sum_{s=3,4} \left\{ t_s\, v_s(0) \prod_{j \neq s} [1 - t_j\, v_j(0)] \right\} = t_3\, v_3(0)[1 - t_4\, v_4(0)] + t_3\, v_3(0)[1 - t_4\, v_4(0)]$$

$$= .40(1 - .20) + .20(1 - .40) \;=\; .32 + .12 \;=\; .44$$

Fault tree vulnerability: Component A OR Component B

$$\Pr(OR) = 1 - \prod_{s=A,B}[1 - t_s v_s(0)] = 1 - (1 - .40)(1 - .44) = 1 - (.60)(.56) = .664$$

The initial risk associated with the fault tree in this example is simply the sum of initial risks across all hazards:

$$R = \sum_{i=1}^{4} t_i v_i(0) c_i = .80c_1 + .50c_2 + .40(1 - .20)c_3 + .20(1 - .40)c_4$$

The last two terms use the equation for Pr(XOR) to obtain initial vulnerability. Thus, .40(1-.20) is the initial probability of hazard h_3, but not h_4, and .20(1-.40) is the initial probability of h_4, but not h_3. The dependency of h_3 on h_4 and h_4 on h_3 will pose a problem for optimal resource allocation, as described below.

If we further assume consequences are 100, 200, 300, and 400, respectively, the initial risk is:

$$R = .80(100) + .50(200) + \{.40(1 - .20)(300)\} + \{.20(1 - .40)\}(400)$$
$$= 80 + 100 + 96 + 48 = 324$$

Resource allocation is MBRA's second innovation with respect to fault tree analysis. What is the best allocation of a fixed budget P to hazards such that vulnerability is reduce and risk is minimized? Optimal allocation of P to n hazards in a fault tree with only AND and OR gates is directly calculated from the risk equation:

$$\min_{p_i} R = \sum_{i}^{n} t_i v(\mathrm{p}_i)\, c_i$$

$$v(\mathrm{p}_i) = v_i(0)e^{-\gamma_i \mathrm{p}_i}; \quad \gamma_i = -\ln[\frac{v_i(\infty)}{v_i(0)}]/\mathrm{p}_i(\infty)$$

subj

$$P = \sum_{i}^{n} p_i\, ; \mathrm{p_i} \geq 0$$

Employing a single Lagrange multiplier to transform this constrained optimization of n variables into an unconstrained optimization of (n+1) variables, differentiating with respect to each variable, setting the resulting equations to zero, and solving the (n+1) simultaneous equations produces closed-form solutions for p_i.

$$\ln(\lambda) = \frac{\sum_{i=1}^{n}\{\frac{\ln(\gamma_i t_i v_i(0))}{\gamma_i} - P\}}{\sum_{i=1}^{n}\frac{1}{\gamma_i}}$$

$$p_i = \frac{\ln(\gamma_i t_i v_i(0))}{\gamma_i} - \ln(\lambda)$$

These equations work for a fault tree with AND and OR gates, and provides a starting point for optimizing a fault tree with one or more XOR gates. For example, assuming the fault tree of Figure 1 contains an OR in place of the XOR, and using the cost data as follows: P = 20, assume for i = 1,2, 3, 4:

$$p_1 = 5.00, p_2 = 6.20, p_3 = 5.55, p_4 = 3.25$$

This allocation reduces vulnerability as follows:

$$h_1 : 80\% \rightarrow 19.9\%$$
$$h_2 : 50\% \rightarrow 12.0\%$$
$$h_3 : 40\% \rightarrow 12.6\%$$
$$h_4 : 20\% \rightarrow 12.8\%$$

Risk is reduced from 324 to 121.56 and fault tree vulnerability drops to 24.0% from 66.4%.

This calculation was obtained by an iterative algorithm that starts by allocating P = 20 to the fault tree containing OR gates in place of XOR gates. It uses the closed-form solution given above as a starting point:

$$p_1 = 4.38$$
$$p_2 = 5.44$$
$$p_3 = 6.41$$
$$p_4 = 3.77$$

In a second phase, MBRA repeatedly selects a donor hazard and a target hazard at random, moves a random amount from the donor to the target, recalculates risk, and tests the exchange to determine if risk has increased or not. If risk has increased, the exchange is canceled. Otherwise, the exchange is kept. The iterative algorithm is:

XOR Fault Tree Allocation Algorithm

1.0. Temporarily replace XOR gates with OR gates and find an initial allocation of P to $[p_1, p_2, \ldots p_n]$. Revert to the original fault tree with its XOR gates.

2.0. Repeat for t = 1, 2,…

2.1. Select a donor hazard at random, DONOR.

2.2. Select a target hazard at random, TARGET

2.3. Save p_{DONOR} and p_{TARGET}, as SAVED_DONOR, SAVED_TARGET.

2.4. Calculate and save fault tree risk, RISK.

2.4. Add a random amount, STEP, to p_{TARGET}

2.5. Subtract STEP from p_{DONOR}

2.6. Recalculate risk, NEW_RISK

2.7. if NEW_RISK > RISK

2.7.1. Restore p_{DONOR} and p_{TARGET} to saved SAVED_DONOR and SAVED_TARGET

2.8. Until there is no additional change to RISK, e.g.

$$\left| \frac{RISK(t) - RISK(t-1)}{RISK(t)} \right| < \varepsilon$$

Fault tree optimization on trees of 20 to 30 hazards takes a matter of seconds on a desktop computer using this algorithm. Note that convergence to a true minimum risk is not guaranteed, but as a practical matter, allocations are very close. Additionally note that the STEP size is a random number, typically calculated as a random fraction of the difference between allocation p_i and p_i (∞).

Network Risk

The simple PRA approach does not model an interdependent system of components and assets; what we define as a *network*. Lewis and Al-Mannai used network theory to model critical infrastructure such as water, power, energy, transportation, and telecommunication systems as networks. Their model represents a system of assets as an abstract graph, G={N, E, f}, where N is a set of n assets called nodes; E is a set of m relationships or connections called links, and f: N x E is a mapping of links to node pairs. G defines a network or system of i = 1, 2, … n+m assets – each with its own vulnerability v_i and consequence c_i.

In terms of PRA, the individual risk associated with asset i (node or link) is simply $z_i = v_i\ c_i$, and so the total risk contribution over n+m assets is simply $Z = \sum z_i$. Unfortunately, this simple definition of risk ignores the synergistic and interdependency effects of a network. For example, it ignores the fact that a network node with many connections is more important to the overall functionality of the system than a node with only one connection.

Albert and Barabasi (2000) studied the vulnerability of structured networks (versus random networks) in terms of their ability to remain connected. They found that degree sequence $g = \{g_1, g_2, \ldots, g_n\}$ of network structure makes a major difference in the survivability of a network. If the network is random, its degree sequence distribution will obey a binomial distribution.

In a scale-free network the distribution follows a power law. Scale-free networks can be protected by focusing on high-degreed hub nodes at the expense of less connected nodes. Albert and Barabasi assumed all nodes and links are of equal value, however, which rarely occurs in the real world. Lewis introduced weights, representing consequences, on nodes and links, and extended the Albert-Barabasi model (Lewis, 2006).

Lewis (2006) generalized the Albert-Barabasi result by introducing consequences; each node and link is assigned a consequence value in addition to a network science weight such as degree, betweeness, or height within a directed network. But Lewis assumed a linear relationship between allocation and reduction of vulnerability, which is unrealistic. In addition, Lewis did not provide a closed-form solution to the minimum risk allocation problem as in this paper.

Al Mannai and Lewis (2007) give closed-form solutions to the problem of allocating a fixed budget to nodes and links such that risk is minimized, where network risk is defined as:

$$Z = \sum_{i=1}^{n+m} g_i v_i c_i$$

where

g_i = degree of node if asset i is a node, and 1 if asset i is a link.
v_i = probability of failure, if attacked.
c_i = damage/consequence if asset i fails.
n = number of nodes
m = number of links

They compared linear and non-linear models of vulnerability versus allocation and showed that overall network risk is minimized by allocating more resources to nodes and links with higher values of the product: $g_i v_i c_i / v_i(\infty)$, where $v_i(\infty)$ is the cost to eliminate vulnerability of node i. The Al Mannai-Lewis model considers risk from the defender's point of view, and ignores the attacker.

Major (2002) proposed an attacker-defender model using a zero sum game-theoretic approach. Major showed how to find an optimal allocation of resources for both attacker and defender, but considered only a single asset or target. Major defined expected loss or risk (EL) as a function of both attacker and defender resources, and the value of the asset.

$$\text{EL} = \textstyle\sum_i V_i\, p(V_i, A_i, D_i), \text{ and } p(V_i, A_i, D_i) = \left\{ e^{\frac{-A_i D_i}{\sqrt{V_i}}} \right\} \left\{ \frac{A_i^2}{A_i^2 + V_i} \right\}$$

Vulnerability, $p(V_i, A_i, D_i)$ is the probability of total destruction of target i; V_i, the target value; A_i, the resource assigned to target i by the attacker, and D_i, the defender allocation to defend the asset. Instead of this model, we study two other models -- linear and exponential -- that represent defender and attacker vulnerability.

Powers (2005) extended Major's (2002) probability model for a single asset by allowing simultaneous attacks on multiple assets. Powers introduced a sophisticated attacker-defender model where the defender wants to minimize the attacker's payoff and the attacker maximizes the defender's payoff. We employ a similar technique to find optimal resource allocation to a network of connected assets, nodes and links.

Powell (2006) presented a basic game-theoretic framework for allocating defensive resources against long-term threats. Resources are allocated to harden sites, reduce vulnerabilities, and make the sites less attractive and difficult to attack. Optimally, the defender will allocate resources to minimize the attacker's payoff, and conversely, the attacker will allocate resources to maximize the defender's payoff. We employ the same min-max strategy in the following model, which extends the defender-attacker model to a network of assets rather than a single asset.

MODEL-BASED RISK ANALYSIS (MBRA)

PRA provides the basis for an attacker-defender model that, when combined with network analysis, comprehensively models system-wide risk. We call this the model-based risk analysis (MBRA) model because the system being analyzed is modeled as a network where each node and link has an associated PRA risk. Additionally, the MBRA risk model uses simulation to obtain a risk and resilience profile for the entire network. The central model and mathematics for resource allocation is given without proof, below. For derivations, see (Al Mannai 2007).

Network risk is defined differently than fault tree risk, because network risk incorporates a network property such as node degree, betweeness, or height. Node degree is equal to the number of links connecting the node to other nodes. Betweeness is the number of paths passing from all nodes to all other nodes, along shortest paths, going through a node or link. Height is the number of hops from sink nodes to source nodes in a directed network. These properties are normalized by dividing them by the largest value among all nodes and links. For example, node degree is normalized to [0,1] by dividing all node degrees by the maximum degree across all nodes. Similarly, betweeness, and height are normalized on [0,1]. Two or more network properties may be combined by multiplying their normalized properties. The combined normalized property g_i, in turn, normalizes risk:

$$z_i = g_i t_i v_i c_i; \quad 0 \le g_i \le 1$$

In addition, MBRA employs a modified Stackleberg optimization to allocate attacker resources to threat with the intent to maximize risk, and to allocate defender resources to vulnerability and consequence with the intent to minimize risk.

This is an extension of the attacker-defender model first proposed by (Major 2002), (Powell 2006), and (Powers 2005).

The objective of the attacker is to strategically "buy up" threat where it will do the most damage, while the objective of the defender is to "buy down" vulnerability and consequence where it reduces risk as much as possible. MBRA iterates between threat allocation and prevention and response allocation until reaching an equilibrium between threat maximization and risk minimization.

The central model of network risk in MBRA is:

$$Z = \sum_{i}^{n+m} g_i\, t(a_i) v(p_i)\, c(r_i)$$

where for each asset (node or link) i:

a_i: investment to increase threat

p_i: investment to reduce vulnerability

r_i: investment to reduce consequence

g_i : weight corresponding to one or more user-selected network science parameters,

$t(a_i)$: threat function defining the probability of an attack

$$t(a_i) = t_i(0); \quad T = 0; \quad t_i(0)\, given$$
$$= 1 - e^{-\alpha_i a_i}; \quad \alpha_i = \frac{-\ln[1 - t_i(\infty)]}{a_i(\infty)}; \quad T > 0; \quad t_i(\infty)\ and\ a_i(\infty)\ given$$

$v(p_i)$: vulnerability function defining the probability of destruction, if attacked

$$v(p_i) = v_i(0)e^{-\gamma_i p_i}; \quad \gamma_i = \frac{-\ln[\frac{v_i(\infty)}{v_i(0)}]}{p_i(\infty)}$$

$c(r_i)$: consequence function defining damages from a successful attack

$$c(r_i) = c_i(0)e^{-\beta_i r_i}; \quad \beta_i = \frac{-\ln[\frac{c_i(\infty)}{c_i(0)}]}{c_i(\infty)}; \quad c_i(0), c_i(\infty) \textit{ given as inputs}$$

n: number of nodes
m: number of links
The network risk function above is constrained by budgets:

$T = \sum_i^{n+m} a_i$; $a_i \geq 0$: T is the attacker's budget, given.

$p_i(\infty)$; $p_i \geq 0$: P is the defender's prevention budget, given

$R = \sum_i^{n+m} r_i$; $r_i \geq 0$: R is the defender's response budget, given.

Figure 2 shows how inputs are used to calibrate these exponentials using estimates of threat, vulnerability, and consequence for given investments. Functions $t(a_i)$, $v(p_i)$, and $c(r_i)$ model the increase in t, and reduction in v and c, given investments a_i, p_i, and r_i, respectively.

Al Mannai and Lewis (Al Mannai 2007) give closed form solutions for risk minimization when functions $t(a_i)$ and $c(r_i)$ are constants and function $v(p_i)$ is linear or exponential.

Figure 3 illustrates a simple network as it appears in the MBRA edit window. Note how an optional map placed in the background facilitates laying out a network over a region of the earth.

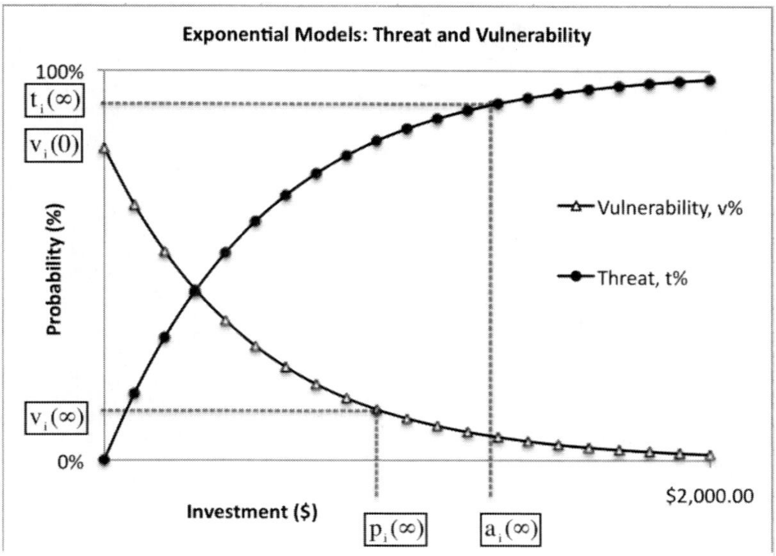

Figure 2(a). Exponential functions used by MBRA to model threat and vulnerability. The slope of each exponential is established by input values shown in boxes.

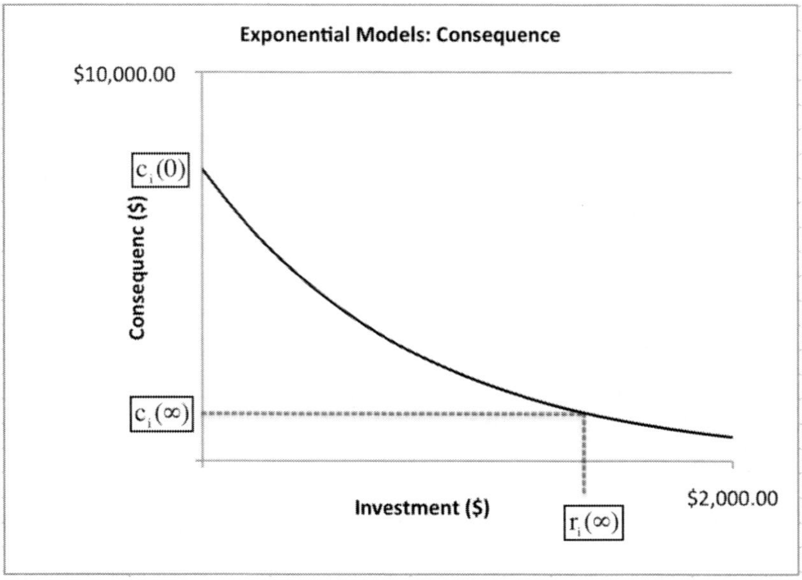

Figure 2(b). Exponential function used by MBRA to model consequence. The slope is established by input values shown in boxes.

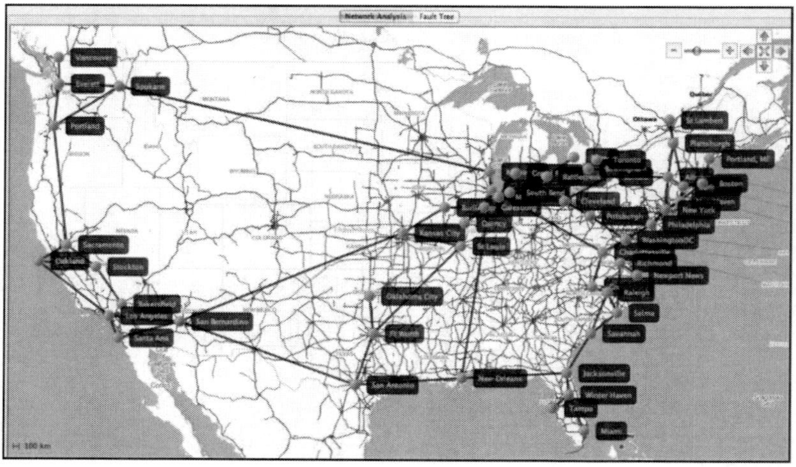

Figure 3. Network model of major Amtrak routes in the US. Nodes are stations and links are railways. Data for each node and link is entered through dialogs exactly like the one shown in Figure 1(b).

This is a useful feature for modeling physical systems such as pipelines, waterways, power lines, roads, etc. However, the map does not impact the analysis in any way, and may be turned off.

NETWORK RESOURCE ALLOCATION

MBRA simulates an attacker-defender Stackleberg game where an adversary attempts to maximize network risk, and a defender attempts to minimize network risk, Z. Both attacker and defender have a limited and fixed resource to apply toward optimization. However, the defender's resources are divided into two parts: a prevention budget used to reduce vulnerability and a response budget used to reduce consequence. Threat, vulnerability, and consequence are all modeled as exponential functions, see Figure 2.

The objective function for resource allocation is:

$$\underset{pi,r_i}{Min}\left\{\underset{a_i}{Max}\ Z\right\}$$

Vulnerability and consequence both decline exponentially with investment. So minimization is applied sequentially. On the other hand, the adversary applies the threat budget all at once, knowing the allocation of the defender. Conversely, the defender may observe the allocation of threat against nodes and links of the network, so the defender attempts to minimize risk, given the attacker's allocation strategy. This process of allocation, counter-allocation, and re-allocation is repeated until a stalemate is reached whereby additional improvement is impossible. This is a Stackleberg game with sequential defender allocations (prevention and response) prior to each attacker allocation. The optimization is Stackleberg and sequential as follows:

Stackleberg/Sequential Resource Allocation Algorithm

1.0. Initially Z_0 is risk with no investment, and no iterations.
1.0. Repeat for t = 1,2…
1.1. Allocate prevention budget P to minimize with respect to vulnerability: $Z_0 \rightarrow Z_1$
1.2. Allocate response budget R to minimize with respect to consequence: $Z_1 \rightarrow Z_2$
1.3. Allocate attack budget T to maximize with respect to threat: $Z_2 \rightarrow Z_3(t)$
1.4. Until there is no change, e.g. $\left| \frac{Z_3(t) - Z_3(t-1)}{Z_3(t)} \right| < \varepsilon$

Sequential allocation algorithms are obtained in the same manner as before with fault tree optimization. We use a Lagrange multiplier to remove constraints, differentiate the resulting objective function with respect to prevention, response, and attacker allocations, set to zero, and solve for investment amounts. The resulting closed-form solutions are shown here. Note that each solution is expressed in terms of other solutions. The circular dependency of a solution on previous solutions is why a sequential algorithm is used.

Allocate Prevention Budget

$$\ln(\lambda_p) = \frac{\sum_{i=1}^{n+m} \frac{\ln[g_i t(a_i) v_i(0) c(r_i)]}{\gamma_i} - P}{\sum_{i=1}^{n+m} 1/\gamma_i}$$

$$p_i = \frac{\ln[g_i t(a_i) v_i(0) c(r_i)] - \ln(\gamma_i)}{\gamma_i}$$

Allocate Response Budget

$$\ln(\lambda_c) = \frac{\sum_{i=1}^{n+m} \frac{\ln[g_i t(a_i) v(p_i) c_i(0)]}{\beta_i} - R}{\sum_{i=1}^{n+m} 1/\beta_i}$$

$$r_i = \frac{\ln[g_i t(a_i) v(p_i) c_i(0)] - \ln(\lambda_c)}{\beta_i}$$

Allocate Attack Budget

$$\ln(\lambda_t) = \frac{\sum_{i=1}^{n+m} \frac{\ln[\alpha_i g_i v(p_i) c(p_i)]}{\alpha_i} - T}{\sum_{i=1}^{n+m} 1/\alpha_i}$$

$$a_i = \frac{\ln[\alpha_i g_i v(p_i) c(r_i)] - \ln(\gamma_t)}{\alpha_i}$$

Note that each of the three allocation equations above is expressed in terms of the other two. Prevention allocation p_i is a function of r_i and a_i; r_i is a function of p_i and a_i; and a_i is a function of p_i and r_i. [If any budget is zero, the corresponding allocation is zero – a calculation that can be made without the allocation equation.

However, in the case of threat, a budget of $T = 0$ means the input values of t_i are used in place of one]. This explains why the Stackleberg optimization is performed sequentially: first the p_i and r_i are set to zero.

The first values of a_i are allocated based on initial p_i and r_i values. Then, new values of p_i are calculated based on previous r_i and a_i. Subsequently, new values of r_i are calculated based on previous a_i and p_i, etc. The process is iterated until there is no change in risk.

Threat, vulnerability, and consequence are driven by budgets. In extreme cases the only inputs needed to perform a complete risk analysis are estimates of consequences and the cost of prevention and response. Rational values of threat and vulnerability can be calculated from such meager input data.

That is, assuming a rational actor model, MBRA is able to calculate risk, make optimal allocations, and output expected values of threat and vulnerability. [Threat and vulnerability are typically very difficult to estimate.

Thus, MBRA addresses one of the major obstacles to risk analysis by calculating t and v values instead of requiring that the user enter them].

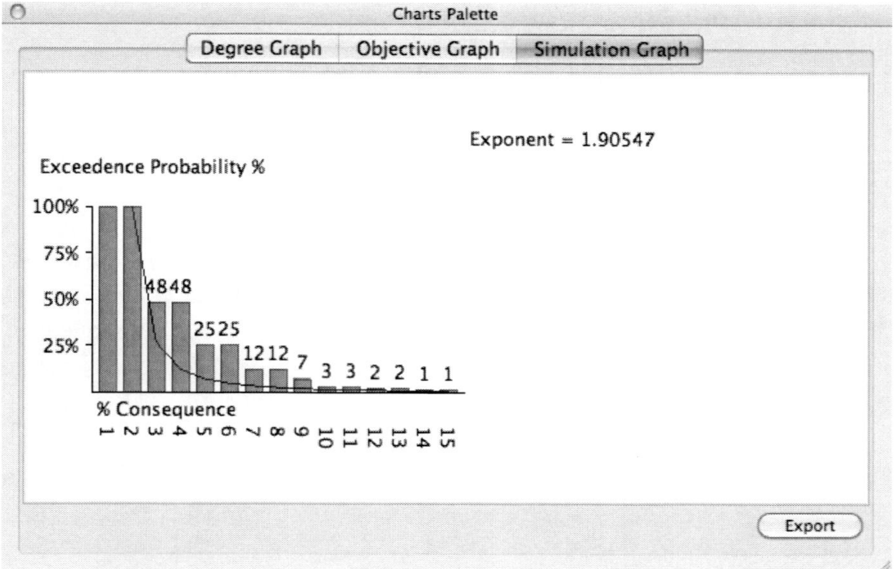

Figure 4. Exceedence probability of cascade episodes obtained by simulation of the Amtrak network shown in Figure 3. The exceedence probability fits a power law with exponent q = 1.9.

SIMULATION

MBRA uses simulation in an unusual way – to calculate exceedence probabilities, and in turn, probably maximum loss, PML. Exceedence probability EP(x) is the probability that an event of size x or larger will occur with in a system – not just an asset.

This is ideal for network analysis because network failures are systemic.

Grossi and Kunreuther explain how exceedence probability is applied to risk assessment using probable maximum loss, PML, in place of expected value (Grossi 2005).

PML risk is the expected loss due to a hazard of size x or larger. It is a worst-case estimate of consequence rather than an average-case estimate.

$$PMLRisk = EP(c) \bullet c$$
$$c = consequence$$
$$EP(c) = \int_{c}^{\infty} \Pr(x)dx$$
$$\Pr(x) = \frac{freq(x)}{\int_{0}^{\infty} freq(x)dx}$$

MBRA uses discrete-event simulation, hence the results are discrete-valued. A fault episode is initiated by selecting a single node or link to fail. Then the consequence of the episode is recorded and analyzed. Consequences from k episodes are recorded as a sequence c_1, c_2,...c_k, converted into a frequency histogram and then normalized into a probability distribution. Finally, the probability distribution is converted into a discrete exceedence probability, EP(c_i).

$$Given: c_1, c_2 \ldots c_k$$
$$freq(c_i): number\ of\ times\ c_i\ occurs$$
$$\Pr(c_i) = \frac{freq(c_i)}{\sum_{i=1}^{k} freq(c_i)}$$
$$EP(c_{n-i}) = EP(c_{n-i+1}) + \Pr(c_{n-i}); i = 1, 2, \ldots n-1$$

Cascade Risk

Cascades are commonly occurring fault episodes in many complex systems such as power grids, nuclear power plants, telecommunicaiton systems, and Internet exploits. The idea is simple: starting with a failed node or link propagate the fault to adjacent nodes with probability of $t_i v_i$. Sum up the consequences and save in c_j. Repeat episodes until k consequence totals have been recorded. Calculate and display the exceedence probability as shown in Figure 4. Each episode starts by selecting a node or link at random and marking it as failed. The selection is random, but proportional to the $t_i v_i$ value of the node/link. This approximates what one would expect: that node/link failure rates should be proportional to likelihood of failure.

Similarly, propagation to adjacent nodes is proportional to the $t_i v_i$ value of the adjacent nodes/links.

Flow Risk

Flow risk is very similar, however, it requires a directed network. That is, all links must be unidirectional. Flow simulation simulates the flow of the objective function value through the network from source nodes to sink nodes. A source node is automatically defined as a node with zero incoming links. Sink nodes are defined as nodes with zero outgoing links. MBRA starts flow simulation by calculating the total values of objective functions across all sink nodes. This is the baseline output of the network. Then MBRA selects one node or link at random (proportional to the $t_i v_i$ value of the node/link), and marks it as failed. This means the node/link will disable all of its outputs to outgoing links. Finally, MBRA re-calculates output of the network by simulating network flow without the marked node or link. This produces a consequence as follows:

$$Given\ flows: c_0, c_1, c_2 \ldots c_k; where\ c_0 = nonfailure\ flow$$

$$Calculate\ decreases\ in\ total\ flows: fc_i = \frac{c_0 - c_i}{c_i}$$

$$freq(fc_i): number\ of\ times\ fc_i\ occurs$$

$$\Pr(fc_i) = \frac{freq(fc_i)}{\sum_{i=1}^{k} freq(fc_i)}$$

$$EP(fc_{n-i}) = EP(fc_{n-i+1}) + \Pr(fc_{n-i}); i = 1, 2, \ldots n-1$$

Network flow is simulated by an iterative process whereby the output of each node at time t is distributed to all outgoing links proportional to their objective function values set by the user. The proportionality is calculated as the ratio of node objective value versus total output from node i:

$$fraction_i = \frac{obj_i}{\sum_{j \in outgoing} obj_j}$$

Inputs from time t are used to update outputs at time (t+1). This is repeated until there are no changes in outputs. To perform this computation, MBRA forms a capacity matrix C, from the objective function selected by the user; a fraction matrix F, from the proportionality calculation above, and a state matrix S, defined as the flow through each node. In matrix form:

$$S_t = \min\left\{C, F^T S_{t-1}\right\}; t = 1, 2, \ldots$$

$$F^T = transform(F)$$

$$S = \begin{bmatrix} s_1 \\ s_2 \\ \ldots \\ s_n \end{bmatrix}$$

$$F = \begin{bmatrix} f_{1,1} & f_{1,2} & \cdots & f_{1,n} \\ f_{2,1} & \cdots & & \\ \cdots & & & \\ & & & f_{n,n} \end{bmatrix}$$

$$C = \begin{bmatrix} c_{1,1} & c_{1,2} & \cdots & c_{1,n} \\ c_{2,1} & \cdots & & \\ \cdots & & & \\ & & & c_{n,n} \end{bmatrix}$$

The state equation contains the min{} function because nodes are not allowed to overflow their maximum capacities. The capacity matrix incorporates the network's topological connections or network structure as well as the maximum values of the objective function selected by the user. Finally, the calculations terminate when there is no longer a change in S, or n iterations have occurred.

Exceedence Probability

Figure 4 shows the exceedence probability curve obtained by simulating cascade failures in the Amtrak network shown in Figure 3. Note that exceedence probability is typically a power law. In Figure 4, MBRA fits a power law power(x) = x^{-q} to the simulation data, with q = 1.9.

REFERENCES

Al Mannai 2007. Al Mannai, W. and Lewis, T., "Minimizing Network Risk with Application to Critical Infrastructure Protection," *Journal of Information Warfare*, Aug. 2007, 6 (2): 52-68.

Albert 2000. Réka Albert, Hawoong Jeong, Albert-Lászlő Barabási, The Internet's Achilles' Heel: Error and attack tolerance of complex networks, *Nature*, 406: pp. 200, (2000), *http://www.neci.nec.com/~giles/barabasi/ nature00.p*

Arquilla 2005. Arquilla, John and David Ronfeldt, (ed), "Networks and Netwars: The Future of Terror, Crime, and Militancy", RAND Corp. The RAND Corporation, 1776 Main St. P.O. Box 2138, Santa Monica, CA 90407-2138 *https://ssl.rand.org/cgi-bin/cart.pl,* 380pp., ISBN: 0-8330-3030-2

Barabasi 2002. Barabasi, Albert-Laszlo, *Linked: How Everything Is Connected To Everything Else and What It Means for Business, Science, and Everyday Life*, Perseus Publishing, 11 Cambridge Center, Cambridge. MA, 02142, 280 pp., 2002. ISBN 0-7382-0667-9.

Bernoulli 1954. Daniel Bernoulli, Exposition of a New Theory on the Measurement of Risk, *Econometrica* 22 (1954) pp23-36.[Translated from Latin into English by Dr. Louise Sommer, The American University, Washington, D. C., from "Specimen Theoriae Novae de Mensura Sortis," Commentarii Academiae Scientiarum Imperialis Petropolitanae, Tomus V, 1738].

Buchanan 2002. Buchanan, Mark, "Nexus: Small Worlds and the Groundbreaking *Theory of Networks*," W. W. Norton, 500 Fifth Ave. NY, NY, 10110, 235 pp., 2002. ISBN 0-393-32442-7 pdk.

Final Report 2003, "Final Report on the August 14, 2003 Blackout in the United States and Canada: Causes and Recommendations", April 2004. *US-Canada Power System Outage Task Force*, pp. 12, (January 12, 2004) at https://reports.energy.gov.

Grossi 2005. Patricia Grossi and Howard Kunreuther, *Catastrophe Modeling: A New Approach to Managing Risk, Springer*, New York, NY. (2005): 245pp.

Lewis 2006. Lewis, T. G., *"Critical Infrastructure Protection in Homeland Security: Defending a Networked Nation,"* John Wiley and Sons, 111 River Street, Hoboken, NJ, 07030-5774 486pp., 2006, ISBN: 0-471-78628-4.

Major 2002. Major, J., "Advanced Techniques for Modeling Terrorism Risk," *Journal of Risk Finance*, Fall 2002, volume 4, issue 1.

Perrow 1999. Charles Perrow, *Normal Accidents*, Princeton University Press, Princeton, New Jersey, (1999): 450pp.

Powell 2006. Powell, R., "Defending Strategic Terrorists Over the Long Run: A Basic Approach to Resource Allocation," *Institute of Governmental Studies (University of California, Berkeley)*, paper WP2006-34.

Powers 2005. Powers, M. R. and Shen, Z., "Colonel Blotto in the War on Terror: Implications for Event Frequency," *Fox School Working Paper.* Temple University.

Rasmussen 1975. Rasmussen, Norman C., et al. (1975-10). "Reactor safety study. An assessment of accident risks in U. S. commercial nuclear power plants. Executive Summary." WASH-1400 (NUREG-75/014). Rockville, MD, US: US Nuclear Regulatory Commission.

Strogatz 2001. Strogatz, Steven, "Exploring Complex Networks", *Nature*, (410)8, March 2001, *www.nature.com.*

Watts 2003. Watts, Duncan, J., *Six Degrees: The Science of a Connected Age*, W. W. Norton, 500 Fifth Ave. NY, NY, 10110, 374 pp., 2003. ISBN 0-393-32542-3 pbk.

In: Network Topologies
Editor: Casandra Rendell

ISBN: 978-1-62618-180-9

Chapter 5

BIOMEDICAL AND CLINICAL NETWORKS OF CANCER

***Ertugrul Dalkic*[1,2] *and Christina Chan**[*1,2,3,4,5]

[1]Cellular and Molecular Biology Lab,
Department of Chemical Engineering and Materials Science,
Michigan State University, East Lansing, MI, US
[2]Cell and Molecular Biology Program,
Michigan State University, East Lansing, MI, US
[3]Department of Chemical Engineering and Materials Science,
Michigan State University, East Lansing, MI, US
[4]Department of Biochemistry and Molecular Biology,
Michigan State University, East Lansing, MI, US
[5]Department of Computer Science and Engineering,
Michigan State University, East Lansing, MI, US

ABSTRACT

Cancer is a large class of diseases. Systems level analysis of complex diseases such as cancer requires the analysis of relationships between different types of clinical as well as molecular data. Chemotherapy, in addition to radiotherapy and surgery, is used to treat cancers. However, cancer drugs mostly target general processes, i.e. DNA synthesis, toinhibit cell division. A recent trend has been to use targeted

[*] To whom correspondence should be addressed (email: krischan@egr.msu.edu).

chemotherapy, to more selectively treat specific types of cancer. Systems science is the analysis of the global organization of the relationships between groups, i.e. the different types of diseases and drugs, and is in contrast to traditional approaches of analyzing individual parts of a system. Such analyses have provided an understanding of the combined clinical and molecular aspects of diseases such as cancer. Networks of diseases, drugs, and genes showed unexpected connectivity for particular members, correlations between the connectivity with epidemiological features, etc. Network analysis of clinical, medical and molecular data is a promising branch of systems science.

CLINICAL THERAPIES OF CANCER

Cancer is a large class of different diseases which results from the deregulation of cell growth [1]. There are more than 200 types of cancers, comprising of different organs, tissues including lung cancer, breast cancer, colorectal cancer, etc. [2]. It is the second leading cause of death in the United States. The United States leads the world with the highest number of cancer deaths [2]. Approximately 12 million Americans alive in 2008 had a history of cancer, some of whom are currently cancer free. Around one and a half million Americans are expected to be diagnosed with cancer and approximately half a million Americans are expected to die of cancer in 2012 [1]. These statistics support the need for more clinical research as well as developmentof more effective therapies for cancer.

Clinical treatment of cancer includes chemotherapy, radiotherapy, surgery, transplantation, etc. [3]. Chemotherapy is the use of drugs to slow down or prevent the growth of the cancer cells. The goal of chemotherapy is to reduce the cancer by destroying the tumor cells, slowing the growth of the cancer, controlling the growth of the cancer by preventing the spread of the tumor cells, or easing the symptoms (i.e., pain) associated with the growth of the cancer. Classical cancer drugs include i) mitotic poisons or tubulin inhibitors, such as Paclitaxel, which disrupt the formation of mitotic spindle and thereby progression of the cell cycle, ii) anti-metabolites,such as Cytarabine, which are nucleic acid base analogues that are incorporated into the DNA to stop DNA synthesis, iii) topoisomerase inhibitors,such as Epipodophyliotoxins, which are inhibitors of topoisomerase I or topoisomerase II that prevent DNA replication, and iv) cell cycle phase-independent agents, such as Cisplatin, Doxorubicin, etc., which alkylate the guanine bases to crosslink two strands of DNA, or intercalate between the neighbor bases, or locate in the external grooves of the DNA to disrupt the double helix topology [2]. There are side-effects of chemotherapy such as nausea, diarrhea, hair loss, fatigue, secondary cancers, etc., because the non-cancer cells could also be damaged [4]. Managing the dosage and the delivery of the cancer drugs is also an important aspect of the chemotherapy. There is only a small range between toxicity and efficacy of most cytotoxic cancer drugs, which necessitates caution in determining the dosage of the chemotherapy drugs [4]. Radiation therapy is the use of high energy radiation, such as X-rays, gamma rays, etc., to damage the DNA and thereby destroy the tumor cells [3]. There also could be side effects to radiotherapy as the normal cells are damaged with this therapy. Finally, different types of cancer therapies have been combined; for instance, surgery can be followed by radiotherapy or chemotherapy [3]. A challenge of cancer therapy is to avoid the recurrence of tumor growth [2]. For this reason, tumor markers are monitored to check the progression of the disease.

Cancer drugs generally target the DNA structure, function, and synthesis to halt or stop cell division. Such drugs are not likely to selectively kill a specific tumor or to differentiate between normal and tumor cells. In recent years however, there has been progress in targeted cancer therapies, which are drugs that target molecular markers of a specific cancer [3]. For example, the SRC family protein tyrosine kinases are downstream factors of several cell membrane receptors and relay the signal to proliferation-promoting genes. Dasatinib is a direct inhibitor of the SRC family tyrosine kinases and is approved for the treatment of leukemia. Targeted therapies are more promising, as they are more specific to the tumor cells, and thus are less likely to harm other (i.e. normal) cells. The success of the cancer treatment is dependent on the cause of the cancer, external factors such as smoking or internal factors such as mutation, and the type of cancer. For example, cancers caused by smoking could be prevented [3]. Also, if a certain mutation is found in a patient, targeted chemotherapy for that molecule could be more effective. For example, the cancer drug Vemurafenib inhibits the activated mutated form of the BRAF serine/threonine kinase protein in melanoma patients and found to be successful [3]. In recent years, treating cancers with combination of chemotherapy drugs has gained significant attention. When a single protein is targeted by a drug, the cell can use a redundant pathway to continue proliferation. In combination chemotherapy, different drugs targeting different molecules or pathways couldmitigate the impact of redundant pathway activity; as well as reduce toxicity due to the lower doses of the drugs used [2].

The majority of the current cancer drugs target general mechanisms, such as DNA synthesis, which do not provide specificity and also result in drug resistance, therefore clinical trials for new targeted cancer therapies (i.e. drugs) are of high priority [2]. For this purpose, both basic and translational research are needed. While basic cancer research aims to find novel molecular mechanisms involved in the cancer, translational research aims to improve cancer therapy by capitalizing on the basic research findings [2]. Basic research investigates the modified genes in cell lines through functional studies, such as analyzing how the upregulated genes impact proliferation or apoptosis. On the other hand, translational research investigates the potential of clinical use of drugs that target the genes in question, i.e., screening drugs that inhibit the upregulated gene product. If the results are encouraging, it is followed up with *in vivo* studies in transgenic mouse models and then finally clinical trials [2].

Clinical trials of novel drugs provide evidence for their efficacy, risks and optimal use of the drugs in patients [5]. Phase 1 clinical trials are performed on small groups of individuals to evaluate the pharmacodynamics, pharmacokinetics, and dose toxicity of the drug [2]. Phase 2 clinical trials are performed to assess the safety and efficacy of the drug. Phase 3 clinical trials are performed on a large group with similar goals and also to provide comparisons with previously approved drugs. The US Food and Drug Administration (FDA) regulates the approval and labeling of the drugs by evaluating the safety, efficacy, and security of the use of cancer drugs in patient [6]. Since basic molecular studies are essential to the development of cancer drugs, it would behoove us to analyze the relationship between the results obtained through basic cancer research with those obtained clinically. For example, if a common molecular mechanism is found for two different cancer types, this raises the question of whether these cancers also share clinical drug trials or FDA drug approvals. Systems science and network analysis can address these types of question to provide an idea of whether the molecular research findings are reflected in the cancer therapies that are developed.

Networks of Cancer

In recent years the biomedical and clinical sciences have embraced systems level research that ismore quantitative, integrative, and predictive. Systems science is an interdisciplinary field that focuses on the principles of abstract organization rather than specific properties and produces models to describe the global characteristics or governing principles [7]. Analyzing part of a system in isolation from the rest cannot provide a comprehensive understanding of its properties and thus cannot readily yield predictions of its behavior. This is significant for biomedical research since the interactions between genes, proteins, cells, tissues, and organs are crucial for proper functioning of the organism. In other words, biomedical systems cannot be completely understood by analyzing the properties of the molecules in isolation, but rather is better understood by analyzing the relationships between the molecules [7]. For example, the identification of feedback mechanisms in biomedical system is an observation obtained through a systems level analysis. Systems medicine investigates the regulatory interactions in the clinical systems, such as feedback and feed-forward loops between the patients, clinicians, literature knowledge and data from both basic and translational research [7]. The relationships between different types of disease, drugs, and clinical trials are also important as are the relationships between different molecules. Analyzing the links between diseases provides a system level view of the relationships between the different diseases [8]. These studies assume that two diseases that share a common molecular mechanism, such as a mutation target, could have a common origin. This is supported by the finding that similar disease phenotypes, such as different cancer types, cluster together when they are linked by mutation targets [9]. These network analyses of disease networks provided new insights into comorbidities, which are secondary diseases that occur concurrently with the primary disease [8]. A patient with a particular disease was shown to be more likely to develop a secondary disease that is a neighbor in the disease network [8]. For example, hemolytic anemia is a complication of diabetes mellitus and these two diseases have a common molecular target gene associated with them which is bisphosphoglyceratemutase. Such analysis of disease networks provides a systems level understanding of disease relationships as compared with the traditional classification of human diseases that neglects this higher level of connectivity of diseases. For example, there might be a mutation in a gene which can cause a primary disease and the interacting neighbor of that specifically mutated gene can lead to secondary disease phenotypes, thereby linking these diseases to each other [8].

Network level analysis has been performed for approved drugs and their corresponding diseases [10]. From a global topological analysis of the drug and disease associations certain emergent properties were uncovered of the FDA approved drug therapies, i.e., most diseases are connected to each other by at most through 3 connections. The study revealed a high compactness of the network as well as the presence of drug hubs that connect many diseases to each other [10]. One of the aims of new drug design is to have multiple diseases as targets of a drug, so that multiple related diseases could be prevented by a single treatment. These are the hub drugs in this network level study. Thus new drug designs could focus on increasing such hub drugs in the future.

Network analysis of diseases based on their mutated metabolic enzyme profile, showed a higher likelihood of comorbidity between diseases which are linked in the network [11]. Therefore, two diseases which share molecular mechanisms, such as metabolic regulation, could have a higher likelihood of having a comorbidity relationship between them; such that one could be the secondary disease of the other. Interestingly, diseases with a high number of connections in this disease network had higher number of cases and death rates as compared to other less connected diseases [11]. These results suggest that connectivity of the diseases provides significant information about the disease statistics (in terms of morbidity) and epidemiology (connectivity of primary diseases to secondary diseases). This type of analysis was also performed for cancers alone in order to understand the relationships between the different types of cancers, e.g. whether cancers that share mutation targets are treated with the same drugs. The FDA and clinical trial data were analyzed for such relationships [12]. To analyze for cancer-drug associations (relationships), two networks were generated, a FDA and a clinical trial network. In the FDA network the nodes consist of the different types of cancers and the connections represent a drug that is FDA approved to treat two cancers (Figure 1). An example of a FDA approved drug shared between two cancers is illustrated in Figure 1. Porfimer is FDA approved to treat both esophagus and lung cancer and thereby links these two cancers in a FDA cancer network. Similarly in a clinical trial network the nodes consist of the different types of cancers and the connections represent a drug that is being tested in clinical trials to treat two cancers (Figure 2). Letrozole is a drug used in clinical trials to treat both breast and ovarian cancers and thereby links these two cancers in a clinical cancer network. From these cancer networks the weighted connections were calculated as the ratio of the number of drugs shared by a cancer with other cancers to the number of total drugs available to treat that particular cancer. Then, the weighted degree value of each cancer was calculated as the sum of the weighted connections. The cancers with the highest number of FDA approvals and clinical trials are leukemia, lung cancer, lymphoma, and breast cancer; however, only breast and lung cancers have high and significant weighted degree values in the FDA cancer network [12]. This means that breast and lung cancers share more of their approved drugs with other cancers.

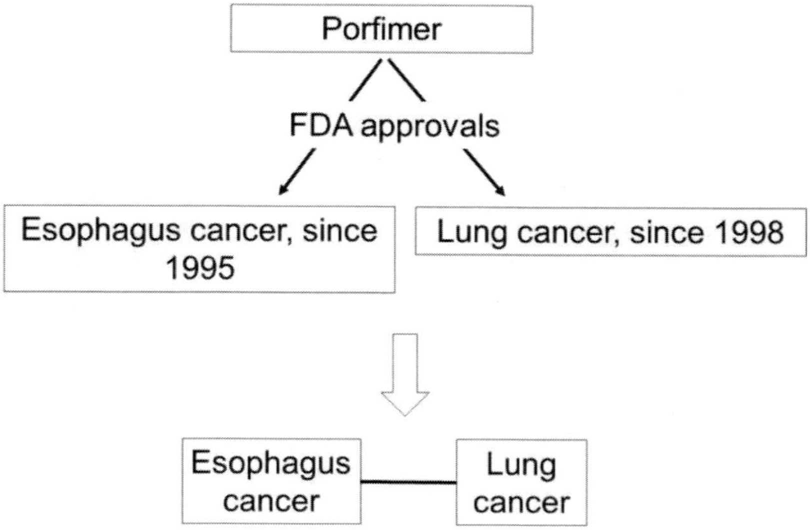

Figure 1. Porfimer was approved for esophagus cancer in 1995 and for lung cancer in 1998. These associations are represented as a link between esophagus and lung cancers based on FDA approved drugs that are shared.

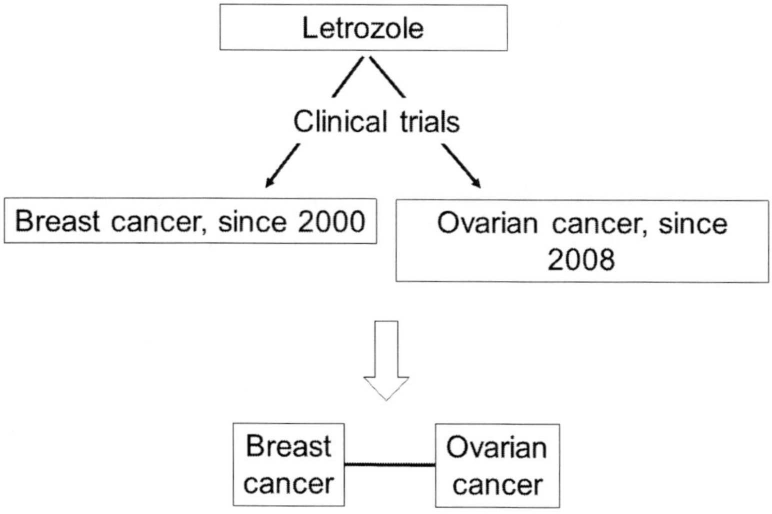

Figure 2. Letrozole was used in completed clinical trials for breast cancer in 2000 and for ovarian cancer in 2008. These associations are represented as a link between breast and ovarian cancers based on drugs that are shared in clinical trials.

In contrast, leukemia and lymphoma also have high number of drug approvals and trials but do not share many of their drugs (approved to treat them) with other cancers, unlike breast and lung cancers. This systems analysis of cancer-drugs associations found that the distribution (number) of the drugs shared among all the different cancers is not the same as the distribution (number) of the drugs approved for the different cancers. Thus network analysis can produced novel information that would not be readily obtained otherwise. It was further found through this analysis that cancer drug approvals and clinical drug trials are correlated to each other. In other words, a cancer with a high number of drug approvals is more likely to have a high number of clinical drug trials, which is to be expected. However, the FDA cancer network, which is based on the FDA approved drugs shared by the various cancers, is very different from the clinical trial cancer network, which is based on the allocation of drugs in clinical trials. In other words, most cancer pairs that are strongly connected in the clinical trial network arenot in the FDA cancer network [12]. This further supports that network level analysis could provide information, from a different perspective, on the distribution of drugs for each cancer. This analysis could provide guidance for future FDA approvals and clinical trial decisions. For example, the connections in both the FDA and clinical trial cancer networks hint to a relationship between the cancers. Thus, the neighboring cancers in these networks could be related to each other and their relationships could befurtherexplored in the future.

Network analysis coupled with epidemiological data could provide information on potential trends associated with the different cancers. From such analysis, it was found that the cancers with the lowest survival rates; pancreatic, liver and esophagus cancers have very few FDA approved drugs as compared with the other cancers [12]. In contrast, lung and breast cancers, which have many more incidences or cases, are highly connected to other cancersthrough the sharingof many drugs with other cancers. However, it is notable that these two cancers do not share many mutation targets with these other cancers for which they share drug therapies [12]. This suggests that they are not related to the other cancers with respect to the molecular mechanisms involved but nevertheless they share similar FDA approved drugsused to treat them. On the other hand, colorectal, ovarian and brain cancers share similar molecular mechanisms according to the mutation-target based network but do not share drugs approved to treat them[12]. These network analysis results could guide future trials and approvals of drugs for colorectal, ovarian, and brain cancers, namely, network analysis could help determine whether the already approved drugs from another cancer could be used to treat these cancers based upon their similarity in their molecular mechanism.

Overall, cancer is a complex disease that includes many different types associated with various tissues and stages. Previous cancer drug therapy focused on drugs that target generic processes, while more recent trend has been to develop drugs for specific targets. Network level analysis of drugs, diseases, and genes, could help identify novel candidates for drug design, as well as monitor and guide current biomedical research on cancer. The analysis of current FDA approved drugs suggests that the cancer-drug associations do not correlate with the molecular based relationships. This is because a majority of the currently approved cancer drugs do not target specific mechanisms, but rather focus on general processes and factors, which can be shared by many cancers. In the future, as the development of specific, targeted cancer therapies increase, similar analysis could be performed to assess for changes in potential correlations between the clinical and molecular information among the cancers as compared with the current state.

REFERENCES

[1] American Cancer Society.Cancer Facts and Figures. Atlanta: American Cancer Society. 2012.

[2] Rachel, A. Cancer Chemotherapy: Basic Science to the Clinic. Wiley-Blackwell, Hoboken, NJ, USA, 2009.

[3] http://www.cancer.gov/cancertopics/treatment/types-of-treatment.

[4] Priestman, T. Cancer Chemotherapy in Clinical Practice. Springer, New York, USA, 2012.

[5] *http://www.clinicaltrials.gov.*

[6] *http://www.fda.gov.*

[7] Deisboeck TS, Kresh JY. Complex Systems Science in Biomedicine. Springer, New York, USA, 2006.

[8] Barabasi AL, Gulbahce N, Loscalzo J. Network medicine: a network-based approach to human disease. *Nat. Rev. Genet.* 2011, 12(1):56-68.

[9] Goh KI, Cusick ME, Valle D, Childs B, Vidal M, Barabasi AL.The human disease network. *Proc. Natl. Acad. Sci. USA*. 2007, 104(21):8685-90.

[10] Nacher JC, Schwartz JM. A global view of drug-therapy interactions. *BMC Pharmacol.* 2008, 8:5.

[11] Lee DS, Park J, Kay KA, Christakis NA, Oltvai ZN, Barabási AL.The implications of human metabolic network topology for disease comorbidity. *Proc. Natl. Acad. Sci. USA*. 2008, 105(29):9880-5.

[12] Dalkic E, Wang X, Wright N, Chan C. Cancer-drug associations: a complex system. *PLoS One*. 2010, 5(4):e10031.

INDEX

C

D

E

I

K

L

O

P

Q

R

T

U

V

W

Y

Z